Physics of Complex Systems

This book analyses the physics of complex systems to elaborate the problems encountered in teaching and research. Inspired by the of Kurt Gödel (including his incompleteness theorems), it considers the concept of time, the idea of models and the concept of complexity before trying to assess the state of physics in general.

Using both general and practical examples, the idea of information is discussed, emphasizing its physical interpretation, and debating ideas in depth, using examples and evidence to provide detailed considerations on the topics. Based on the authors' own research on these topics, this book puts forward the idea that the application of information measures can provide new results in the study of complex systems.

Helpful for those already familiar with the concepts who wish to deepen their critical understanding, *Physics of Complex Systems* will be extremely valuable both for people that are already involved in complex systems and also for readers beginning their journey into the subject. This work will encourage readers to follow and continue these ideas, enabling them to investigate the various topics further.

Physics of Complex Systems
Discovery in the Age of Gödel
First Edition

Dragutin T. Mihailović
Darko Kapor
Siniša Crvenković
Anja Mihailović

CRC Press is an imprint of the
Taylor & Francis Group, an **informa** business

Front cover image: By the kind permission of his granddaughter Emilija Nikolić-Đorić:"The clover on the front cover was carried inside a pocket Bible by soldier Petar Nikolić throughout the Great War (1914–1918). Here it symbolizes the continuity of life and tradition as one tiny plant in the immense universe."

First edition published 2024
by CRC Press
2385 NW Executive Center Dr, Suite 320, Boca Raton, FL 33431

and by CRC Press
4 Park Square, Milton Park, Abingdon, Oxon, OX14 4RN

CRC Press is an imprint of Taylor & Francis Group, LLC

© 2024 Dragutin T. Mihailović, Darko Kapor, Siniša Crvenković and Anja Mihailović

Reasonable efforts have been made to publish reliable data and information, but the author and publisher cannot assume responsibility for the validity of all materials or the consequences of their use. The authors and publishers have attempted to trace the copyright holders of all material reproduced in this publication and apologize to copyright holders if permission to publish in this form has not been obtained. If any copyright material has not been acknowledged please write and let us know so we may rectify in any future reprint.

Except as permitted under U.S. Copyright Law, no part of this book may be reprinted, reproduced, transmitted, or utilized in any form by any electronic, mechanical, or other means, now known or hereafter invented, including photocopying, microfilming, and recording, or in any information storage or retrieval system, without written permission from the publishers.

For permission to photocopy or use material electronically from this work, access www.copyright.com or contact the Copyright Clearance Center, Inc. (CCC), 222 Rosewood Drive, Danvers, MA 01923, 978-750-8400. For works that are not available on CCC please contact mpkbookspermissions@tandf.co.uk

Trademark notice: Product or corporate names may be trademarks or registered trademarks and are used only for identification and explanation without intent to infringe.

ISBN: 978-1-032-22801-3 (hbk)
ISBN: 978-1-032-24418-1 (pbk)
ISBN: 978-1-003-27857-3 (ebk)

DOI: 10.1201/9781003278573

Typeset in Palatino
by SPi Technologies India Pvt Ltd (Straive)

Dedicated to all those who taught me and whose stories I listened to.
　　　　　　　　　　　　　　　　　Dragutin T. Mihailović

Dedicated to my grandchildren: Tara, Ema and Filip
　　　　　　　　　　　　　　　　　　　　　Darko Kapor

Dedicated to my granddaughter Višnja
　　　　　　　　　　　　　　　　　Siniša Crvenković

Dedicated to my parents
　　　　　　　　　　　　　　　　　Anja Mihailović

Contents

Preface xi
About the Authors xv

1. **Prolegomenon** 1
   1.1 The Generality of Physics 1
   1.2 Physics: A Crisis that Has Been Lasting for a Century!
       Is that Really So? 4
   1.3 Complex Systems in Physics 9
   1.4 Physics and Mathematics Walk Together Along a
       Narrow Path 11
   References 15

2. **Gödel's Incompleteness Theorems and Physics** 19
   2.1 Gödel's Biography and the Historical Background of
       Incompleteness Theorems 19
   2.2 An Informal Proof of Gödel's Incompleteness
       Theorems of Formal Arithmetic 22
   2.3 Gödel's Incompleteness Theorems as a Metaphor. Real
       Possibilities and Misrepresentation in their Applications 26
   2.4 Gödel's Work in Physical Problems and Computer Science 28
   References 31

3. **Time in Physics** 33
   3.1 Time in Philosophy and Physics: Beyond Gödel's Time 33
   3.2 Does the Quantum of Time Exist? 36
   3.3 Continuous and Discrete Time 38
   3.4 Time in Complex Systems 41
   References 45

4. **Are Model and Theory Synonymous in Physics? Between
   Epistemology and Practice** 49
   4.1 Some Background Concepts and Epistemology 49
   4.2 Choice in Model Building 52
   4.3 The Discrete Versus Continuous Dichotomy: Time and
       Space in Model Building 54
   4.4 The Predictability of Complex Systems. Lyapunov and
       Kolmogorov Time 56

		4.5 Chaos in Environmental Interfaces in Climate Models..................58
		References 61

5. **How Can We Assimilate Hitherto Inaccessible Information?**.............65
   5.1 The Physicality, Abstractness, and Concept of Information..........65
   5.2 The Metaphysics of Chance (Probability)..67
   5.3 Shannon Information. The Triangle of the
        Relationships Between Energy, Matter, and Information...............70
   5.4 Rare Events in Complex Systems: What Information
        Can Be Derived From Them?...72
   5.5 Information in Complex Systems..75
   References 77

6. **Kolmogorov and Change Complexity and Their
   Applications to Physical Complex Systems**..81
   6.1 Kolmogorov Complexity: An Incomputable Measure
        and Lempel-Ziv Algorithm...81
   6.2 Change Complexity: A Measure that Detects Change...................83
   6.3 Kolmogorov Complexity in the Analysis of the LIGO
        Signals and Bell's Experiments..87
   6.4 Change Complexity in the Search for Patterns
        in River Flows..92
   References 94

7. **The Separation of Scales in Complex Systems: "Breaking"
   Point at the Time Scale**..97
   7.1 The Generalization of Scaling in Gödel's World.
        Scaling in Phase Transitions and Critical Phenomena...................97
   7.2 The Separation of Scales and Capabilities of the
        Renormalization Group 101
   7.3 A Phase Transition Model Example: The Longevity of
        the Heisenberg Model 105
   7.4 Complexity and Time Scale: The "Breaking" Point
        with an Experimental Example...108
   References 111

8. **The Representation of the Randomness and Complexity of
   Turbulent Flows**..117
   8.1 The Randomness of Turbulence in Fluids.....................................117
   8.2 The Representation of the Randomness and
        Complexity of Turbulent Flows with Kolmogorov
        Complexity...121
   8.3 The Complexity of Coherent Structures in the
        Turbulent Mixing Layer 124

	8.4	Information Measures Describing the River Flow as a Complex Natural Fluid System 128
	References 132	
9.	**The Physics of Complex Systems and Art** 135	
	9.1	An Attempt to Grasp the Complexity of the Human Brain 135
	9.2	The Dualism Between Science and Art 139
	9.3	Perception: Change Complexity in Psychology 142
	9.4	Entropy, Change Complexity, and Kolmogorov Complexity in Observing Differences in Painting 146
	References 150	
10.	**The Modeling of Complex Biophysical Systems** 153	
	10.1	The Role of Physics in the Modeling of the Human Body's Complex Systems 153
	10.2	The Stability of the Synchronization of Intercellular Communication in the Tissue with the Closed Contour Arrangement of Cells 159
	10.3	The Instability of the Synchronization of Intercellular Communication in the Tissue with a Closed Contour Arrangement of Cells: A Potential Trigger for Autoimmune Disorders 162
	10.4	The Search for Information in Brain Disorders 166
	References 171	

Appendix A 173

Appendix B 175

Index 179

Preface

The inspiration for the book was found in the problems that the authors encountered in teaching and research. These problems required explanations that could not be adequately given in lectures or research papers. Therefore, the book aims to present our considerations and experience to a broad audience and invite readers to investigate various topics by themselves. The main subject of the book is the physics of complex systems. There are two important guidelines in this study. One is the work of Kurt Gödel, whose daring approach through reasoning opposite to accumulated experience and mainstream inspired many researchers to make breakthroughs. The other is describing and analyzing complex systems with information measures. To be more specific: Why did we choose this title and the specific content of the book? The first reason is encompassed by John von Neumann's comments:

> ... there have been within the experience of people now living at least three serious crises... There have been two such crises in physics—namely, the conceptual soul searching connected with the discovery of relativity and the conceptual difficulties connected with discoveries in quantum theory... The third crisis was in mathematics. It was a very serious conceptual crisis, dealing with rigor and the proper way to carry out a correct mathematical proof. In view of the earlier notions of the absolute rigor of mathematics, it is surprising that such a thing could have happened, and even more surprising that it could have happened in these latter days when miracles are not supposed to take place. Yet it did happen.

The second reason was Gödel's original proof of incompleteness theorems with a paradoxical assertion that is true but not provable within the usual formalizations of number theory. In our opinion, his contributions to the foundations of mathematics, philosophy, and physics unite the past, the present, and the far future into "Gödel's age." Gödel was a unique scientist and mathematician whose discovery was like "Deus ex machina." He "demolished" the old house and built a new one. The physics of complex systems was a fruitful field for traveling through "Gödel's age."

The Serbian filmmaker Emir Kusturica states, "Today, everyone writes but no one reads." The same can be said for the scientific and mathematical community. We tried to avoid following this already obvious trend. From the very beginning, the authors were sure about what this book was *not* intended to be. First, it was not supposed to be a textbook, and scholarly explanations were avoided. However, they sometimes appeared when we thought most readers might be unfamiliar with all the concepts and terminology. Next, it was not considered to be an encyclopedia, even though a wide range of topics was covered in the book. We used quotations and cited literature (around 10–20 percent of the size of the chapter) to keep the sizes of the chapters within a tolerable limit.

The creation of this book was an interactive process. There was first interaction between the authors to determine the main subjects of the book and its style. At the next stage, the authors exchanged ideas on every particular subject, which was a very vivid interaction indeed. Once a consensus was achieved, the material was written down, and further steps were defined. The fourth author took care of the consistency and readability of the book. The interaction with the editors, who ensured that the book maintained its proper form and remained within the projected scope, also contributed to the text's final form. The choice of examples was dictated by the authors' experience, particularly by their research. Luckily, our research interests are broad, meaning that many different materials can be found in the book. Actually, some of the examples belong to as-yet unpublished research results. One way of testing whether we were on the right track was to put the material on the internet (arXiv) when it was completed. The feedback proved very helpful.

When going through this book, the reader may come across some (or even many) oddities that can be either criticized (with or without reason) or left aside without finding a way to elaborate on them because the authors were not more convincing. The authors are aware of this, but in some places, these oddities arose due to the authors' intentions. One of those oddities is the fact that the whole book, both in its conception and in its examples, relies on only two information measures: (1) Kolmogorov complexity and (2) change complexity (Aksentijevic-Gibson complexity). Why were these chosen? Anthropologically speaking, people are not used to thinking abstractly, needing confirmed rules and facts for further induction. Accordingly, most scientists are focused on, and seek confirmation of, what happened while ignoring what could have happened. Hence, physicists initiate the "trap of induction" and search for easily comprehended information, not information that is hardly reachable to our minds. They rely on hypotheses and more or less heuristic models as their possible realization. By this approach, we can get results that can be "impressive," particularly in the age of powerful technology that governs the mainstream in physics. However, it seems that an evident step forward in the study of complex systems can be taken if we keep in mind the following points: 1. Each meaningful information is encoded in different complex patterns and is always less complex than maximum complexity (randomness); and 2. Catching the change in a complex system is a crucial step. This is the reason why we selected the information measures mentioned above. One is only a natural but incomputable measure (Kolmogorov complexity), while the other is computable, catching the change. For the end of the short history of this book, we left one detail we noticed when half of the book was already written. In many examples, the difference logistic equation appeared with an unknown but unexpected trait that was completely distant from our intuition. Its appearance reminded us of a picture in which the most magnificent animal on the Earth, a blue whale, emerges from the most powerful part of the Earth and spins back into the mysterious depth of the ocean.

Chapter 1 is a discursive introduction to the book. Here we consider the current state of physics. We perceive the further progress of physics as a synthesis of collective efforts that includes seemingly unrelated fields—physics, mathematics, chemistry, biology, medicine, psychology, and even the arts. In that synergy, physics contributes to the domain of fundamental discoveries analogously to the way entities in a complex system contribute to the system as a whole. We support our standpoint by considering complex systems, the relationship between physics and mathematics, and the increasing power of computational physics. Chapter 2 deals with Kurt Gödel, a great mathematician who initially wanted to study physics. We explain his scientific background by analyzing his biography, give concise proof of Gödel's incompleteness theorems, and point out real possibilities and misconceptions about applying these theorems in physics. In addition, we present his contributions to physics and information science. He wrote two papers inspired by Einstein's theory of relativity, in which he derived a model of the universe allowing time travel. Einstein called this "a great achievement for the theory of relativity." In Chapter 3, we first offer a list of the most important issues under discussion in the philosophical community regarding time. Then we shortly outline the understanding of time in theoretical and computational physics since, in complex systems, it operates concurrently at different scales, runs at multiple rates, and integrates the function of numerous connected systems (complex time). We discuss the consequences of treating time as a continuous or discrete variable and conditions, leading to the possible quantization of time. Chapter 4 is devoted to models and their use in a computer simulation in physical science, which inherently involves many epistemological questions. Although heuristic considerations often overrule the problems of epistemology, it is sometimes necessary to make basic epistemological choices, especially in modeling. Being aware of its weak spots, physicists often ignore the nonlinearity of phenomena and processes by applying the method of linearization. However, if we consider the existing nonlinearities in the object that we model as much as possible, we can recognize the following key points: (1) model choice, (2) continuous time versus discrete time in model building, (3) model predictability (Lyapunov time), and (4) chaos in environmental interfaces in climate models. Chapter 5 discusses information and its relation to physics. We address the following aspects: (1) the physicality and abstractness of information, (2) the concept of information, (3) the metaphysics of chance (probability), (4) information and no event registered, and (5) information in complex systems. We elaborate on two information measures (Kolmogorov complexity and its derivatives and change complexity) that are useful in analyzing various time series in Chapter 6. The examples included here are the search for patterns in the analysis of Bell's experiments, the identification of gravitational waves (LIGO experiment), and environmental fluid flows. At the beginning of Chapter 7, we set one view on the separation of scales in complex systems as a reflection of Gödel's incompleteness theorems. We point out the limits of the renormalization group related to the separation of scales. A need for new mathematics for scaling in complex systems is emphasized. At the end, we give some examples (the Heisenberg model and the "breaking" point on a time scale). Chapter 8 discusses randomness in turbulent flows and its quantification via complexity, and also considers information measures suitable for its description. Finally, we consider the connection of the complexity of coherent structures with turbulent eddies' sizes. We elaborate on the dualism between physics and art in Chapter 9. We emphasize the place of the physics of complex systems in creating an impression about a picture through 1. perception analyzed with change complexity; 2. the recognition of order and disorder with entropy. In Chapter 10, we present some of the contributions of the physics of complex systems to medical science. It is completed in terms of models and approaches that deal with intercellular communication, autoimmune diseases, and brain disorders.

The preparation of the text (both a creative and a technical part) was performed under the permanent supervision of Ms. Betsy Byers, Editorial Assistant for Physics, CRC Routledge, Taylor and Francis, who was there for us whenever we needed advice; we were grateful for her assistance. Prof. Zagorka Lozanov-Crvenković offered us help with the parts of mathematics with which we were unfamiliar and the best way of presenting it, so we owe her a lot. The authors appreciate the contribution of Mr. Miloš Kapor, who produced the artwork (original illustrations). The fourth author, Anja Mihailović, in addition to her remarkable contributions, always carefully and patiently took care to ensure the consistency of the text and language.

About the Authors

Dragutin Mihailović is a retired Professor in Meteorology, Environmental Fluid Mechanics, and Biophysics at the University of Novi Sad (Serbia). He obtained his BSc. in Physics and his MSc. and PhD in Meteorology at the University of Belgrade (Serbia). He was the Visiting Professor or Researcher at a number of institutions in the USA, Netherlands, and Norway. He has more than 200 peer-reviewed scientific papers in international journals and a monograph on subjects related to land-atmosphere processes, air pollution, chemical transport modeling, boundary layer meteorology, physics and modeling of phenomena on environmental interfaces, modeling of complex systems, nonlinear dynamics, and complexity. He was among the editors of six monographs and co-authored a book on the modeling of environmental surfaces (Elsevier). Prof. Mihailović was the principal investigator in the FP6 project and several international projects (the University of Athens, Colorado State University, and the University at Albany-SUNY), a member of the Editorial Board of Environmental Modelling and Software (1992–2010), and a reviewer in several international journals.

Darko Kapor is the retired Professor of Theoretical and Mathematical Physics at the University of Novi Sad (Serbia). He received his BSc. and PhD in Physics at the University of Novi Sad and an MSc. at the University of Belgrade (Serbia). His main research interest is in the area of Theoretical Condensed Matter Physics and was the head of several research projects. Later, he developed an interest in theoretical meteorology. He has authored more than 120 peer-reviewed scientific papers in international journals and chapters in research monographs and has also co-authored a book on environmental surfaces (Elsevier). He has also spent some time visiting laboratories in France, the USA, and Hungary. Prof. Kapor invested much effort in Physics popularization by working with talented pupils and teachers and organizing Physics problem-solving contests. His experience from this work was important while co-authoring textbooks in Physics for elementary and secondary schools.

Siniša Crvenković is a retired Professor of Mathematics, the University of Novi Sad (Serbia). He received his BSc. and PhD degrees in Mathematics at the University of Novi Sad and MSc. at the University of Belgrade (Serbia). In addition to his teaching activities at the Department of Mathematics, the University of Novi Sad, he was a Visiting Professor of Mathematics at the Department of Mathematics, the University of Banja Luka, Bosnia and Herzegovina, for 18 years. His main research interests are Algebra and

Mathematical Logic, Constructive Mathematics, History, and Philosophy of Mathematics. He was the head of several scientific projects and has more than 100 peer-reviewed scientific papers in scientific journals and chapters in research monographs. He was a postdoc IREX grantee in the USA. Prof. Crvenkovic was active in the popularization of Mathematics through his work with talented pupils and also through his involvement in organizing Serbian Math Olympiads. For several years, he was the President of the Board of Trustees of the Serbian Mathematical Society.

Anja Mihailović completed her MSc. in Applied Mathematics at the Faculty of Sciences, the University of Novi Sad (Serbia). She was the External Associate of the Center for Meteorology and Environmental Modelling at the University of Novi Sad. Her scientific research includes 17 peer-reviewed papers primarily focused on complexity, information measures, and their applications.

1

Prolegomenon

1.1 The Generality of Physics

In considering the current progress in science, technology, and its trends, one might be assured that discoveries and advances in complex systems theory and applications will dominate the twenty-first century. This branch of study will improve our understanding of physical, biological, ecological, and social phenomena. Accelerated research, owing to the help of technology and a vast amount of data, has led to the emergence of new fundamental theoretical questions in science. These new problems cannot be solved with the help of traditional methods; rather, they require the development of a new interdisciplinary science of complex systems. Physics will be one of the most important parts of this new approach since solutions to many problems in other sciences rely on physical laws. The contribution of physics to modern science should be defined carefully because its universality and widespread applications resulted in the generality of physics. The generality of physics is the idea that phenomena (irrespective of their origins and nature) fall under the domain of physics and should be explained by its laws. This is especially accepted in biology, but to what extent can physics explain complex biological systems?

According to many dictionaries, physics is concerned with nature and the properties of matter and energy. This includes mechanics, heat, light and other radiation, sound, electricity, magnetism, and the structure of a substance. The concept of matter is often equated with a material substance that, together with energy, forms the basis of all objective phenomena. *Inert matter* comprises any matter that is not a seed, including broken seeds, sterile florets, chaff, fungus bodies, and stones. A remarkable part of the physical community takes it for granted that inertia is intrinsic to matter. However, according to Ernst Mach, a German physicist, inertia only refers to the matter in the universe. By contrast, living organisms are structured, and contain specialized, coordinated parts. They are made up of either one cell (unicellular organisms) or multiple cells (multicellular organisms), which are the fundamental units of life. Physics and biology appear to be pioneers in exploring these two complex worlds. Nowadays, many scientists consider physics general, and biology is regarded as merely particular, but this assertion is an improper one. It can be said that physics incorporates all matter in nature, including organisms, since they belong to material nature. From such an ideal perspective, biology is simply a part of physics (when we say physics, we mean *contemporary* physics). Still, physical laws apply to the limited number and types of material systems; physics is inherently inadequate to accommodate phenomena in biology. Erwin Schrödinger, the author of the book *What Is Life?* [1], was one of the outstanding theoretical physicists of the twentieth century, indeed perhaps of the whole preceding millennium. He regards physics as the ultimate science of material nature and concludes that organisms are repositories of what he calls new physics. Robert Rosen [2] says that Erwin Schrödinger, while permanently asserting the universality of contemporary physics, also points out the complete failure of its laws to explain anything significant about the biosphere and its structure. Albert Einstein also states the same in a letter sent to Leo Szilard, which he describes more vividly: "One can best feel in dealing with living things how primitive physics still is" [3]. By contrast, Jacques Monod [4], three decades after the appearance of Schrödinger's essay, writes, "Biology is marginal because—the living world constituting but a tiny and very 'special' part of the universe—it does not seem likely that the study of living things will ever uncover general laws applicable outside the biosphere." Therefore, this book only addresses the breakthroughs of physics in biology and medicine for which physical processes are obvious and clearly defined.

What is the level of generality of physics or any other scientific or mathematical discipline? We can assume intuitively that the "level of generality" of a theory characterizes the class of situations that the theory can deal with, or which the theory can accommodate. How can something like that be measured? It is illustrative to look at number theory, which comprises many conjectures that no one has ever been able to prove or disprove. The rising question was whether number theory was general enough to explain problems that had arisen. This situation became interesting when Kurt Gödel showed how to represent the assertion about number theory within number theory [5]. Considering things from this perspective, he demonstrated that for any given set of axioms in number theory, there are always propositions similar to theorems that cannot be proved from these axioms. If such a situation already exists in number theory, then we can imagine how difficult it is to consider a similar problem in physics. Gonzalo [6] explains that Stephen Hawking [7] believes that Gödel's incompleteness theorems [2] make the search for the theory of everything impossible. He reasons that because there exist mathematical results that cannot be proved, there must also exist physical results that cannot be proved. Exactly how valid is his reasoning? Opinions on this view have been very divergent. We support the opinion that Stephen Hawking [7] does not claim that results are unprovable in an abstract sense. He assumes that the theory of everything will be a particular finite set of rules and presents an argument that no such set of rules will be sufficient. Gonzalo states that "the final statement that it may not be possible to formulate a theory of the universe in a finite number of statements, which is reminiscent of Gödel's theorem" [6].

The above question is precisely the one raised by reductionism in physics, which is understood as methodological reductionism. It is the attempt to reduce explanations to smaller constituents and explain phenomena as relations between more fundamental entities [8] that connect theories and knowledge within physics. Note that the problems associated with physical reductionism, at least applied naively, are as follows: 1. It misses the emergent properties of a system (reductionism assumes that emergent properties are nothing more than the sum of the reduced properties applied over a very large scale). 2. It is misapplied in biology and other sciences [9]. The question we talk about is an assertion, conjecture, or belief pertaining to the generality of contemporary physics. Unlike Goldbach's conjecture in number theory, conjectures in the physical world are not based on direct evidence. "It is rather indirect (circumstantial) evidence, insofar as evidence is adduced at all. In short, it rests on faith" [2]. This faith was described metaphorically by the Serbian writer Borislav Pekić in the novel *Atlantis* [10]: "The existence of spirits in principle does not contradict any law of physics. It is in contrast to the mind that civilization has modeled on empirical evidence for centuries." Conjectures are based on general experience and are very limited since they are always subject to the so-called "black swan" effect. ("Black swan" is an expression for an event that is very rare, completely unexpected, and unpredictable. It has an enormous impact since it can easily cause the collapse of a complete scientific structure because of conjectures that exclude its existence [11].) At the end of this subchapter, we must stress that everything we said about physical systems was quite general in nature. We will later make a clear distinction between the systems with an infinite or very large number of constituents and those with interfaces compared to the rest of the physical systems.

In order to offer better insight, let us summarize perspectives. The right direction is to list the problems that will be the focus of the scientific community in the twenty-first century. Some of the possible choices, in no particular order, include: (1) whether or not we are alone in the universe; (2) the emission of harmful greenhouse gasses and climate change; (3) the understanding of consciousness; (4) decision-making in an insecure world; (5) the extension of both the maximum and average lifespan; (6) whether the culture is only characteristic of people; (7) managing the Earth's resources; (8) the importance of the internet; (9) the use of master cells in the future; (10) the importance of maintaining biodiversity; (11) the role of reengineering and climate change; and (12) the importance of new vaccines. Let us also mention that, during the preparation of this book, we submitted some early versions of chapters, and the interested reader can look at an extended version of this subchapter [12].

1.2 Physics: A Crisis that Has Been Lasting for a Century! Is that Really So?

Many articles on the crisis in science have been published, beginning with the pioneering paper written by Freistadt [13]. Nonetheless, the clear identification of its origins remains elusive to most commentators [14]. Because of the increasing public impact of the crisis on trust in institutions, authors argue that 1. the crisis in science (hence physics) exists and includes positions and social functions of science (crisis context); 2. the mainstream interpretation of its causes is not sufficient and has to be complemented with insights provided by some researchers who predicted the current dilemma (a lack of root causes); 3. the profound transformation of society and the impact of science on society are, without doubt, induced by this crisis (scientists have also contributed to its creation and had a significant influence on preserving the status quo [the science vs. stake context]); 4. some social mechanisms can be applied to improve this situation, including the essential changes in behavior and social activity of the scientific community (social context).

The crisis and its consequences are reflected in awarding the Nobel Prize. Fortunato [15] points out that the time between publishing research and receiving the Nobel Prize has become longer. This trend is the least present in physiology or medicine and the most present in physics. Before 1940, 11 percent of prizes in physics, 15 percent of those in chemistry, and 24 percent of those in physiology or in medicine were awarded for research that was more than 20 years old. Since 1985, those percentages have risen to 60 percent, 52 percent, and 45 percent, respectively.

The "crisis of physics" implies that everything is affected and that all results are questionable. In our view, this is not the issue, and we state that the crisis exists in physics. What is the crisis in physics? We perceive it as the absence of a fundamental discovery, or, more precisely, the crisis in physics arises and lasts until a new fundamental discovery is made. This expectation is not the expectation of a person who will never appear similar to the one in Samuel Beckett's tragicomedy *Waiting for Godot* [16]. Many modern concepts about the crisis in physics envisage an ending identical to the end of this tragicomedy; we do not see ending but rather a transition. From our standpoint, a fundamental discovery is made by scientists with exceptional thinking abilities who revolutionize physics. We label such progress as *vertical* progress, and it does not matter if it involves a micro, macro, or mega physical world. Other discoveries in physics, the confirmation of existing hypotheses or experimental results providing or needing explanations, are called *horizontal* progress. Our definition of the fundamental discovery needs to be completed since it assumes a large amount of horizontal progress that precedes it. For this reason, we are not confident that the experimental proof of the expanding universe and the quark hypothesis should be treated as fundamental. In our opinion, the three most recent fundamental discoveries in physics are the following: 1. Quantum Theory (Max Planck, 1900) that introduces an individual unit of energy or quantum; the sources are the two outstanding communications of Planck to the Berlin Academy (extraordinary fit of the radiation formula and statistical justification by introduction of the discrete energy elements) collected in the paper by Boya [17]. 2. Theory of special relativity introduced by Albert Einstein in 1905, his annus mirabilis year when he published three papers [18–20] in *Annalen der Physik*. These papers, in which he revolutionized the concepts of space, time, mass, and energy, made major contributions to the foundation of modern physics. 3. Theory of general relativity (Albert Einstein, 1915) concerned with macroscopic behavior and large-scale physical phenomena. Sauer [21] provides the first comprehensive overview of Einstein's final version of the general theory of relativity, published in 1916 after several expositions of the preliminary versions and their latest revisions in 1915. These theories treat the micro world (molecules, atoms, and elementary particles), the macro world (our planet and people), and the mega world or the universe (stars, planets, etc.).

It should be stressed that almost all facts relevant to fundamental discoveries were previously available to scientists. Planck was attempting to explain his extraordinary heuristic fit. Before the theory of general relativity, non-Euclidean geometry was already well known owing to the research of Lobachevsky, Riemann, and Minkowski. The examples of findings before the theory of special relativity are the following: 1. The formulation of the *Lorentz force* [22] as the force acting on charges moving in the magnetic field, which was essential for interactions between the currents. 2. *Lorentz transformations* that are considered constitutive for the theory of special relativity. They were invented by the efforts of Woldemar Voigt, George FitzGerald, and Hendrik Lorentz, while their derivation was described by Heras [23]. 3. The form of *Einstein's equation that relates mass to energy* was known to Henri Poincaré [24], who stated that if one required that the momentum of any particle that is present in an electromagnetic field plus the momentum of the field itself should be conserved together, then Poynting's theorem [25] predicted that the field acted as a "fictitious fluid" with mass satisfying the expression Einstein used later. Unfortunately, the whole theory was based on the concept of ether. The only difference between the scientists mentioned above and Einstein is the way they interpreted available facts. His interpretation made his discovery fundamental and raised physics to a higher level.

What has caused and created this crisis? It seems that the crisis in physics has at least three aspects: 1. *Psychological aspect*. Let us make a digression regarding the term "escape" in the context of positive freedom defined by Fromm [26]. Based on Fromm's definition, it is the capacity for "spontaneous relationship to man and nature, a relationship that connects the individual with the world without eliminating his individuality" [26]. Freedom is also accompanied by loneliness and an inability to exert individual power, and "we use several different techniques to alleviate the anxiety associated with our perception of freedom, including automaton, conformity, authoritarianism, destructiveness, and individuation" [27]. The most common of these mechanisms is conformity. Fromm states that people conform to the larger society and gain power and a sense of belonging by behaving similarly. This power of the masses assists us in not feeling lonely and helpless, but unfortunately, it removes our individuality. How is this phenomenon connected with the crisis in physics? Specifically, the escape from science can be seen as an analogy to the escape from freedom or the "escape from freedom of choice of the research subject" [26]. How is it possible for physicists to escape from physics today? Physicists are willing to believe that cause can be determined, and they accept an obvious solution and declare it the right explanation. On the contrary, they avoid random events that almost always affect and change physical systems. Why? Because physicists find it nearly impossible to deal with random events that evade clear explanations and rules. This ignorance of chance in physical processes leads to scientific conformity and the creation of mainstream in science (hence physics), or the mechanism of escape from physics. Note that the syntagma "mainstream" refers to physicists who have adopted attitudes opposite to Anderson's understanding of physics. Finally, our mind is filled with causality, and we primarily look for the cause instead of accepting a random event. 2. *Epistemological aspect*. a. *Limits of the precision of certainty and strength of Gödel's theorem*. There is no doubt that physics is approaching some limits; the only question is whether they are the limits to how far physics can reach, or its end. To clarify our position, we rely on Gödel's incompleteness theorems. Barrow [28] considers some informal aspects of these theorems and their underlying assumptions and discusses responses to these theorems by people who want to infer something about the completability of physical theories from these theorems. In the same paper, he states "that there is no reason to expect Gödel incompleteness to handicap the search for a description of the laws of Nature, but we do expect it to limit what we can predict about the outcomes of those laws." Stanley Jaki [29] believes that Gödel's theorem prevents us from understanding the cosmos as a necessary truth:

> Clearly then no scientific cosmology, which of necessity must be highly mathematical, can have its proof of consistency within itself as far as mathematics goes. In the absence of such consistency, all mathematical models, all theories of elementary particles, including the theory of truth that the world can only be what it is and nothing else. This is true even if the theory happened to account with perfect accuracy for all phenomena of the physical world known at a particular time.

It constitutes a fundamental barrier to the understanding of the universe:

> It seems that on the strength of Gödel's theorem that the ultimate foundations of the bold symbolic constructions of mathematical physics will remain embedded forever in that deeper level of thinking characterized both by the wisdom and by the haziness of analogies and intuitions. For the speculative physicist this implies that there are limits to the precision of certainty, that even in the pure thinking of theoretical physics there is a boundary...An integral part of this boundary is the scientist himself, as a thinker.
>
> [30]

b. *Limits of decoding information*. Advances in information theory are of critical importance in modern physics. For instance, they are important for detecting the gravitational wave data provided by LIGO (Laser Interferometer Gravitational-Wave Observatory) [31]. However, even if we move the borders of cognition in physics, decoding information remains the problem. Each meaningful information is encoded in different complex patterns and is always less complex than maximum complexity (randomness). In contrast to dealing with easily observed and immediately decodable patterns (e.g., visual input, native language), our cognitive capacity struggles with extracting information from noise. Furthermore, the presence of a pattern (structural information) in noise is a necessary but not sufficient condition for the presence of the meaning (semantic information). Besides, other aspects should be considered: 1. How do we interpret information? 2. Can we search for information outside established concepts? Namely, people are not used to thinking abstractly [11], so they need confirmed rules and facts for further induction. Consequently, most scientists are focused on and seek confirmation of what happened and ignore what could have happened. We lack the dimension of abstract thinking, and we are not even aware of that. Therefore, physicists initiate the "trap of induction" and search for information that is easy to comprehend and not information that is hardly reachable to our minds.

The effect of scientists' emotions and intuition on science and scientific work is worth mentioning. Intuition is the accumulation of experiences that are assimilated unconsciously [32]. Einstein [33] writes an interesting consideration on the importance of intuition: "I believe in intuition and inspiration. Imagination is more important than knowledge. For knowledge is limited, whereas imagination embraces the entire world, stimulating progress, giving birth to evolution. It is, strictly speaking, a real factor in scientific research." Emotions are essential in science, including the justification and discovery of hypotheses [34]. Note that Gödel's work has greatly impacted the speculations about the limitations of the human mind [28]. 3. *Social and economic aspects*. The fact that a major breakthrough in physics has not recently appeared influences the general social attitude towards research. Building large machines (colliders, telescopes) cannot be justified either by pure curiosity or by promising results, but mostly, just as a cosmic flight and Formula 1 racing contribute to the technology, by immediate applications in everyday life. This reasoning has resulted in misinterpretations of scientific results to make them look like something applicable. This aspect is sometimes forgotten, but when it comes to experimental research, it is of great importance.

How do we see the "ending of physics"? In *The End of Physics* [35], David Lindley argues that 1. the theory of everything derived from particle physics will be full of untested and untestable assumptions; 2. if physics is the source of such speculation, it will eventually detach from science and become modern mythology. This will be the end of physics as we know it. Indeed, all big problems in physics have been solved or will soon be. Apparently, no more truly fundamental discoveries, such as quantum mechanics, relativity, the Big Bang, and beyond (cosmology), are left to be made. Therefore, some possibilities for future trends in physics are as follows: 1. It will be aimed at less attractive activities (the research on detailed implications of basic theories, applied problems, etc.). 2. It will be less stimulating and attractive to stellar scientists who can make vertical progress. The Nobel Laureate Philip Anderson presents his observations on physics, its future progression, and directions toward fundamental discoveries [36]. He differentiates two trends in physics—"intensive" and "extensive" research. Intensive research includes fundamental laws, while extensive research comprises the explanation of phenomena in terms of established fundamental laws:

> There are two dimensions to basic research. The frontier of science extends all along a long line from the newest and most modern intensive research, over the extensive research recently spawned by the intensive research of yesterday, to the broad and well-developed web of extensive research activities based on intensive research of past decades.
>
> [36]

He also points out the nature of the hierarchical structure of science and physics. Entirely new properties appear at each level of complexity, and understanding new behavior requires fundamental research. Anderson [36] describes his thoughts on fundamental science through his experience:

> The effectiveness of this message [about intensive and extensive sciences] may be indicated by the fact that I heard it quoted recently by a leader in the field of materials science, who urged the participants at a meeting dedicated to 'fundamental problems in condensed matter physics' to accept that there were few or no such problems and that nothing was left but extensive sciences which he seemed to equate with device engineering.

His comments can be interpreted as a strong criticism of the efforts at replacing science with technology or even as an attempt to transform philosophy into science; philosophy is not science since it employs rational and logical analysis as well as conceptual clarification, while science employs empirical measurements. Similarly, technology cannot replace science. In our opinion, if we consider his conclusions, we can notice continuity and not the end of physics. Our perspective on future developments is similar to Dejan Stojakovic's [37]: "Further progress of physics will require a synthesis of collective efforts in seemingly unrelated fields—physics, mathematics, chemistry, biology, medicine, psychology, and even arts such as literature, painting and music." An additional comment clarifies this sentence more precisely. Physics can contribute to the domain of fundamental discoveries analogously to the way entities in a complex system contribute to a whole system.

1.3 Complex Systems in Physics

A complex system consists of components whose properties and interactions create and influence the whole system. There are many concepts in the science of complex systems, but emergence and complexity are the most fundamental. *Emergence* is a complex system's behavior that is not regulated by the properties of its individual parts but by their interactions and relationships. In other words, emergent behavior occurs only when individual parts interact. Jeffrey Goldstein [38] defines emergence as "the arising of novel and coherent structures, patterns and properties during the process of self-organization in complex systems." There are two notions related to emergence—weak and strong emergence. When interactions between entities at lower levels form new properties at higher levels, then this complex system behavior refers to *weak emergence* (usually called emergence). In the case of weak emergence, at least approximately those low-level processes can be determined. It should be emphasized that "truths concerning that phenomenon are *unexpected* given the principles governing the low-level domain" [39]. Therefore, weak emerging properties are scale-dependent—they are noticeable only if the system is large enough to exhibit a phenomenon. One nice thought on the connection between Gödel's incompleteness theorems and scalability in physics is described by Arthur Seldom in *The Oxford Murders* [40]:

> That exactly the same kind of phenomenon occurred in mathematics, and that everything was, basically, a question of scale. The indeterminable propositions that Gödel had found must correspond to a subatomic world, of infinitesimal magnitudes, invisible to normal mathematics.

The rest consisted in defining the right notion of scale. What I proved, basically, is that if a mathematical question can be formulated within the same "scale" as the axioms, it must belong to mathematicians' usual world and be possible to prove or refute. But if writing it out requires a different scale, then it risks belonging to the world—submerged, infinitesimal, but latent in everything—of what can neither be proved nor refuted.

On the other hand, *strong emergence* is structurally unknown to us—that is, we cannot deduce high-level properties from low-level processes or laws. Undoubtedly, strong emergence involves some processes that we are unable to solve. The only example of strong emergence outside physics is consciousness, while quantum mechanics is a strong candidate for strong emergence within physics. Many researchers have recently challenged the interdependence of consciousness from quantum physics. Hypothetically, we can state that weak emergence and strong emergence are the same with one difference—our capability and incapability to find processes/laws governing the low-level systemic layer. Weak emergence can be simulated, and the crucial point is that the basic entities, for the most part, remain independent after computer simulation. If this does not happen, a new entity with new emergent properties is created. In contrast, there is a broad consensus that strong emergence cannot be simulated nor analyzed. *Complexity* is a nontrivial regularity having its origin inside the system's structure. There is no unique explanation of complexity, and the most general definition is that the system exhibits complexity when its behavior cannot be easily explained by examining its components. Scientists use various complexity measures to assess complexity since we can only obtain information about the complex systems' nature through time series.

We consider a *complex system* (Figure 1.1) as a collection of entities (circles). Each component interacts with others via simple local rules and the

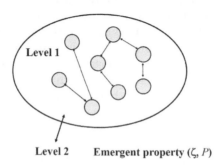

FIGURE 1.1
Toward the definition of a complex system and the concept of the emergent property. (Reproduced by permission from [41].)

possibility of feedback (arrows). When they interact, a new feature arises (level 1). The complex system cannot be decomposed nontrivially into a set of basic entities for which it is the logical sum [2] (to have the character of an emergent phenomenon, this new feature is completely unexpected [level 2]). The new property is characterized by (ζ, P), where ζ represents data, and P is the probability distribution.

Most complicated systems in physics are not complex, but most complex systems in physics are complicated [42]. It is difficult for physicists to comprehend a complex system with various scopes and scales [36]. Paradoxes arise because we can often understand a part of the whole system. Further, we usually 1. embrace a microscopic or macroscopic level but not both simultaneously; 2. analyze the system or the environment but not their mutual interaction. Category theory is a promising choice for modeling a physical system that is hierarchically organized [43]. This was proposed as an alternative to the foundations of mathematics, assuming that morphism is the basis from which everything is built [44].

1.4 Physics and Mathematics Walk Together Along a Narrow Path

In ancient times and even the Middle Ages, mathematics and physics developed together, primarily as parts of philosophy. A typical result of such a relationship is a vector from the study of forces (remember that the name vector is derived from the Latin verb *vehere* meaning "to pull"). It was only later, particularly in the nineteenth century, that they separated completely, and mathematics evolved into an abstract science. Scientists have explored the connection between mathematics and science for more than a century. Almost all papers on this topic begin with "the unreasonable effectiveness of mathematics"—the phrase coined by the physicist Eugene Wigner [45] in 1960. He wonders how "the enormous usefulness of mathematics in the natural sciences is something bordering on the mysterious and that there is no rational explanation for this" [45]. We think that this reasoning was described much earlier in the famous book *The Aim and Structure of Physical Theory* [46] written by the French physicist Pierre Duhem. In the parts "The Aim of Physical Theory" and "Mathematical Deduction and Physical Theory," he explains that 1. theoretical physics is subordinated to metaphysics; 2. mathematical deductions are sometimes helpful in physics, but axioms are usually self-evident truths from which theorems are derived deductively. On the contrary, physical axioms are not strongly self-evident, but physicists proceed with making new conclusions from those axioms, even with mathematical deduction. His book was published when quantum theory and the theory of special relativity were broadly known, while Albert Einstein submitted the theory of general relativity for publication in 1915 (in which he used Riemann's geometry—perhaps the most brilliant use of mathematics in physics).

Mathematics is the science of order, rules, structure, and logical connections. It explores patterns and possible relationships between abstractions irrespective of whether they correspond to phenomena in physics. Gödel's incompleteness theorems basically express that a particular part of mathematics cannot be formalized. This does not necessarily mean formalization would not be a strategy in the study of mathematical systems. These theorems rather indicate that the limits of formalization are the universal strategy. Mathematics provides a theory that scientists can recognize and even use to find similarities between two completely different structures in physics and other sciences. (Noam Chomsky used mathematics to "see" abstract patterns we recognize as grammatical sentences.)

Research in theoretical physics consists of the following steps: 1. Scientists formulate a hypothesis, usually without validating certain conditions. 2. They decide on mathematical methods for their hypothesis. Currently, several problems are related to theoretical physics and its mathematical formalism. Before the choice of a mathematical model, there is a limit imposed by the nature of physics—physics does not include what happens in infinity because this is *beyond* its relative horizon. "The problem of philosophy is to acquire a consistency without losing the infinite into which thought plunges" [47]. (This problem differs from the problem of explaining chaos with the condition of abandoning infinite movements and limitation of speed as condition sine qua non. Chaos is considered as a movement from one determination to another by physicists; from this perspective, chaos introduces disorder and unfastens every consistency in infinity. It seems that the problem is not the determination of chaotic behavior but the infinite speed at which elements are shaped and disappear. Further, physical theories are formalized through mathematical models (Figure 1.2) that can be undecidable; therefore, the problem of *undecidability* exists in physics, especially in condensed matter physics [48]. Last, mathematical models require a priori or complete knowledge about the conditions of their validity.

Alex Harvey emphasizes that a physicist looks for an explanation by mapping physical phenomena onto mathematical objects. This must represent an isomorphism, and then mathematics can be applied [49]. Still, one must be very careful. Physicists often recognize established mathematical formalism, and physics rarely inspires a particular mathematical discovery. Variational calculus or strings might be considered as examples.

Mathematics is related not only to theoretical physics but also to experimental physics (Figure 1.2). Whether it acts as a bridge between them or addition to this classification is still a matter of debate. This also applies to computational physics because it connects two physical branches and is an

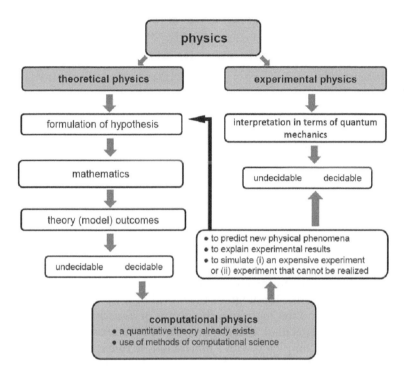

FIGURE 1.2
The relationship between mathematics and different branches of physics.

independent discipline at the same time. Its foundation is theoretical physics, but it additionally implements and combines computer simulations and applied mathematics with physics to solve complex problems. As a result, this branch can help predict new phenomena, explain experimental results, and replace expensive experiments that are difficult to be conducted. The output of a numerical simulation (Figure 1.3) is the behavior of molecules in an ideal gas obtained by Arsenić and Krmar's model [50]. This simple simulation illustrates 1. the spatial distribution of molecules of ideal gas after each collision (which cannot be determined experimentally); 2. the spatial and temporal distribution of ideal gas molecules (which helps discover a new physical phenomenon).

When we state that certain mathematical disciplines might be most helpful in physics in the future, we are aware that this assertion is not unambiguous. At this point, we focus on *category theory* because of its vast potential for future applications in physics. Category theory relates results from two different mathematical fields, algebra and topology—the parts of the foundations of mathematics. It has a central place in contemporary mathematics, theoretical computer science, theoretical physics, and other scientific

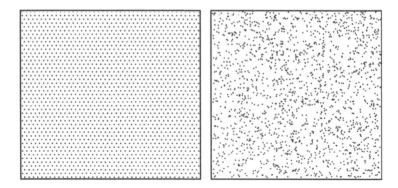

FIGURE 1.3
The spatial distribution of molecules in an ideal gas obtained by a numerical simulation—initial conditions (*left*) and after one million collisions (*right*). (Figure courtesy of Ilija Arsenić and Miodrag Krmar.)

disciplines. Let us imagine an abstract world that includes topological spaces and groups of algebraic topology or the world of unspecified *objects*; this is the world of continuous mapping, the group of *homeomorphisms*, and *morphisms* that are not specified. Our requirement is that those morphisms can be composed. This is completed with a *functor* whose role is to associate objects by morphisms while preserving the identities of structures. Let us suppose that we use category theory to create a model. Then "if we think of functorial images constituting *metaphors* for what they are image, we may ask whether two different such metaphors are the same or not" [2]. The terms "target" and "source" are commonly used to explain metaphor comprehension; for example, in the metaphor "T is S" or "T is like S," T is a target, and S is a source. Generally, the target is moderately unknown, while the source is moderately known [51]. Then we compare relations and functors to assemble relations between them. Here is the point where modeling enters the picture explicitly not (up to now) between the source (S) and target (T) of a single functor but between targets of different functors defined on a common source. This type of comparison is called *natural transformation*, which came first in the history of category, while the rest of category theory came as the reinforcement for it [2]. Thus, we say that we model S in T. This is of crucial importance for physics, particularly for the physics of complex systems, because this formalism provides an opportunity to find *relations* between objects. Usually, physicists perform the following steps that are not always productive work: they deal with objects using selected mathematics and then explore their potential relationship. As the final observation, let us pose a question: How should we look at the relationship between physics and Gödel's incompleteness theorems and his other physics-related ideas? Perhaps the most condensed description is given by John Barrow in the conclusion of his paper [28].

Thus, in conclusion, we find that Gödel's ideas are still provoking new research programmes and unsuspected properties of the worlds of logical and physical reality. His incompleteness theorems should not be a drain on our enthusiasm to seek out and codify the laws of Nature: there is no reason for them to limit that search for the fundamental symmetries of Nature in any significant way. But, by contrast, in situations of sufficient complexity, we do expect to find that Gödel incompleteness places limits on our ability to use those laws to predict the future, carry out specific computations, or build algorithms: incompleteness besets the outcomes of very simple laws of Nature. Finally, if we study universes, then Gödel's impact will always be felt as we try to reconcile the simple local geometry of space and time with the extraordinary possibilities that its exotic global structure allows. Space-time structure defines what can be proved in a universe.

References

[1] E. Schrödinger, *What Is Life?*. Cambridge, U.K.: Cambridge University Press, 1944.
[2] R. Rosen, *Life Itself: A Comprehensive Inquiry into the Nature, Origin, and Fabrication of Life*. New York, NY, USA: Columbia University Press, 1991.
[3] G. J. Klir, *Facets of Systems Science* (IFSR International Series in Systems Science and Systems Engineering 15), 2nd ed. New York; Berlin, Germany; Vienna, Austria: Springer-Verlag, 2001.
[4] J. Monod, *Chance and Necessity*. New York, NY, USA: Knopf, 1971.
[5] K. Gödel, "On formally undecidable propositions of Principia Mathematica and related systems I," (in German), *Mon. Hefte Math.*, vol. 38, no. 1, pp. 173–198, Dec. 1931, doi:10.1007/BF01700692.
[6] J. A. Gonzalo, "Hawking on Gödel and the end of physics," in *Everything Coming Out of Nothing vs. A Finite, Open and Contingent Universe*, Manuel M. Carreira and Julio A. Gonzalo, Eds., Sharjah, UAE: Bentham Sci., 2012, pp. 20–25.
[7] S. W. Hawking, "Gödel and the end of physics" (lecture, Texas A&M University, College Station, TX, Mar. 8, 2002). Available: http://yclept.ucdavis.edu/course/215c.S17/TEX/GodelAndEndOfPhysics.pdf
[8] H. Meyer-Ortmanns, "On the success and limitations of reductionism in physics," in *Why More Is Different: Philosophical Issues in Condensed Matter Physics and Complex Systems* (The Frontiers Collection), B. Falkenburg and M. Morrison, Eds., New York; Berlin, Germany; Vienna, Austria: Springer-Verlag, 2015, pp. 13–39.
[9] G. Longo. "On the borderline: Science and reductionism." *Urbanomic.com*. https://www.urbanomic.com/document/on-the-borderline/ (accessed Nov. 15, 2021).
[10] B. Pekić, *Atlantida*. Belgrade, Serbia: Laguna, 2015.

[11] N. N. Taleb, *Crni labud*. A. Imširović Đorđević and A. Ješić, Trans. Smederevo, Serbia: Heliks, 2016.
[12] D. T. Mihailović, D. Kapor, S. Crvenković, and A. Mihailović, "Physics as the science of the possible: Discovery in the age of Gödel (1.1 Generality of physics)," 2021, *arXiv*:2104.08515.
[13] H. Freistadt, "The crisis in physics," *Sci. Soc.*, vol. 17, no. 3, pp. 211–237, 1953.
[14] A. Saltelli and S. Funtowicz, "What is science's crisis really about?," *Futures*, vol. 91, pp. 5–11, Aug. 2017, doi:10.1016/j.futures.2017.05.010.
[15] S. Fortunato, "Growing time lag threatens Nobels," *Nature*, vol. 508, no. 7495, pp. 186–186, Apr. 2014, doi:10.1038/508186a.
[16] S. Beckett, *Waiting for Godot*. New York, NY, USA: Grove, 1954.
[17] L. J. Boya, "The thermal radiation formula of Planck (1900)," 2004, *arXiv*: physics/0402064.
[18] A. Einstein, "On a heuristic point of view about the creation and conversion of light," (in German), *Ann. Phys.*, vol. 322, no. 6, pp. 132–148, Jan. 1905, doi:10.1002/andp.19053220607.
[19] A. Einstein, "On the electrodynamics of moving bodies," (in German), *Ann. Phys.*, vol. 322, no. 10, pp. 891–921, Jan. 1905, doi:10.1002/andp.19053221004.
[20] A. Einstein, "Does the inertia of a body depend on its energy content?" (in German), *Ann. Phys.*, vol. 323, no. 13, pp. 639–641, Jan. 1905, doi:10.1002/andp.19053231314.
[21] T. Sauer, "Albert Einstein's 1916 review article on general relativity," 2004, *arXiv*: physics/0405066.
[22] F. D. Tombe. "Maxwell's original equations." *Gsjournal.net*. https://www.gsjournal.net/Science-Journals/Essays (accessed Nov. 25, 2021).
[23] R. Heras, "A review of Voigt's transformations in the framework of special relativity," 2014, *arXiv*:1411.2559.
[24] M. H. Poincaré, "On the dynamics of the electron," (in French), *Rend. Circ. Mat. Palermo (2)*, vol. 21, no. 1, pp. 129–175, Dec. 1906, doi:10.1007/BF03013466.
[25] J. H. Poynting, "On the transfer of energy in the electromagnetic field," *Philos. Trans. Roy. Soc.*, vol. 175, pp. 343–361, Dec. 1884, doi:10.1098/rstl.1884.0016.
[26] E. H. Fromm, *Escape from Freedom*. New York, NY, USA: Farrar & Rinehart, 1941.
[27] E. Fromm. "Political psychophilosopher Erich Fromm." *Allpsych.com*. https://allpsych.com/personality-theory/psychodynamic/fromm/ (accessed Dec. 25, 2021).
[28] J. D. Barrow, "Gödel and physics," 2006, *arXiv*: physics/0612253.
[29] S. Jaki, *Cosmos and Creator*. Edinburgh, UK: Scottish Academic Press, 1980.
[30] S. Jaki, *The Relevance of Physics*. Chicago, IL, USA: University of Chicago Press, 1966.
[31] M. G. Kovalsky and A. A. Hnilo, "LIGO series, dimension of embedding and Kolmogorov's complexity," *Astron. Comput.*, vol. 35, Apr. 2021, Art. no. 100465, doi:10.1016/j.ascom.2021.100465.
[32] A. R. Damásio, *Descartes' Error: Emotion, Reason, and the Human Brain*. New York, NY, USA: Putnam, 1994.
[33] A. Einstein and G. B. Shaw, *On Cosmic Religion and Other Opinions and Aphorisms*. New York, NY, USA: Dover, 2009.

[34] P. Thagard, "The passionate scientist: Emotion in scientific cognition," in *The Cognitive Basis of Science*, P. Carruthers, S. Stich, and M. Siegal, Eds., Cambridge, U.K.: Cambridge University Press, 2002, pp. 235–250.

[35] D. Lindley, *The End of Physics: The Myth of a Unified Theory*. New York, NY, USA: Basic Books, 1994.

[36] P. W. Anderson, "More is different: Broken symmetry and the nature of the hierarchical structure of science," *Science*, vol. 177, no. 4047, pp. 393–396, Aug. 1972, doi:10.1126/science.177.4047.393.

[37] D. Stojaković, "Nemoć fizike," *Nova Galaksija*, Jan. 2021, 6.

[38] J. Goldstein, "Emergence as a construct: History and issues," *Emergence*, vol. 1, no. 1, pp. 49–72, 1999, doi:10.1207/s15327000em0101_4.

[39] D. J. Chalmers, "Strong and weak emergence," in *The Re-Emergence of Emergence: The Emergentist Hypothesis from Science to Religion*, P. Clayton and P. Davies, Eds., London, U. K.: Oxford University Press, 2008, pp. 244–254.

[40] G. Martinez, *The Oxford Murders*. San Francisco, CA, USA: MacAdam/Cage, 2005.

[41] M. Fuentes, "Complexity and the emergence of physical properties," *Entropy*, vol. 16, no. 8, pp. 4489–4496, Aug. 2014, doi:10.3390/e16084489.

[42] M. Gell-Mann and S. Lloyd, "Information measures, effective complexity, and total information," *Complexity*, vol. 2, no. 1, pp. 44–52, Sep./Oct. 1996, doi:10.1002/(SICI)1099-0526(199609/10)2:1<44::AID-CPLX10>3.0.CO;2-X.

[43] G. Mack, "Universal dynamics, a unified theory of complex systems. Emergence, life and death," *Commun. Math. Phys.*, vol. 219, no. 1, pp. 141–178, May 2001.

[44] T. Gowers, J. Barrow-Green, and I. Leader, Eds., *The Princeton Companion to Mathematics*. Princeton, NJ, USA: Princeton University Press, 2008.

[45] E. P. Wigner, "The unreasonable effectiveness of mathematics in the natural sciences. Richard Courant lecture in mathematical sciences delivered at New York University, May 11, 1959," *Commun. Pure Appl. Math.*, vol. 13, no. 1, pp. 1–14, Feb. 1960, doi:10.1002/cpa.3160130102.

[46] P. Duhem, *La théorie physique son objet et sa structure*, 2nd ed. Paris, France: Chevalier et Rivière, 1914.

[47] G. Deleuze and F. Guattari, *What Is Philosophy* (European Perspectives: A Series in Social Thought and Cultural Criticism). New York, NY, USA: Columbia University Press, 1994.

[48] T. S. Cubitt, D. Perez-Garcia, and M. M. Wolf, "Undecidability of the spectral gap," *Nature*, vol. 528, no. 7581, pp. 207–211, Dec. 2015, doi:10.1038/nature16059.

[49] A. Harvey, "The reasonable effectiveness of mathematics in the natural sciences," *Gen. Relativ. Gravit.*, vol. 43, no. 12, pp. 3657–3664, Aug. 2011, doi:10.1007/s10714-011-1248-9.

[50] I. Arsenić and M. Krmar, "Numerical simulation of free expansion of an ideal gas," *Mod. Phys. Lett. B*, vol. 35, no. 26, Sep. 2021, Art. no. 2150450, doi:10.1142/S0217984921504509.

[51] M. Fuyama, H. Saigo, and T. Takahashi, "A category theoretic approach to metaphor comprehension: Theory of indeterminate natural transformation," *Biosystems*, vol. 197, Nov. 2020, Art. no. 1042132020, doi:10.1016/j.biosystems.2020.104213.

2
Gödel's Incompleteness Theorems and Physics

2.1 Gödel's Biography and the Historical Background of Incompleteness Theorems

At the dawn of the twentieth century, David Hilbert (1862–1943), who posed Hilbert's second problem, asked for direct proof of the consistency of the theory of numbers, integers, or reals. In 1931 a shocking solution was found by Kurt Gödel, who showed that the consistency of any theory containing the theory of numbers could not be proved in the theory itself. Hence, no theory that must be a foundation for mathematics can justify itself; it must therefore search for justification in an external system. Consistency means that, within the theory, we cannot prove that both sentences A and $\neg A$ are the theorems of the theory. Therefore, no consistent theory containing the theory of numbers can be complete at the same time in the sense that all mathematical truths expressible in its language can be proved in the theory. One of the truths that cannot be proved is its own consistency. This result of Gödel is called *incompleteness theorems*. In this subchapter, we describe Gödel's life and achievements. Almost all mathematical foundations lead to Hilbert—the actual leader of mathematical science from the end of the nineteenth to the first half of the twentieth century. Hence, we first offer a brief description of his work.

Hilbert was born in what was then called Königsberg (Kaliningrad in the Russian Federation). Apart from spending one semester at the University of Heidelberg, he obtained his mathematical training at the University of Königsberg, at which the only full professor of mathematics was Heinrich Weber. Owing to his lectures, Hilbert became aware of number theory and function theory. At that time, he met the young mathematician Hermann Minkowski, with whom he later enjoyed a fruitful cooperation. Weber was later succeeded by the German mathematician Ferdinand Lindemann, who suggested that Hilbert research a problem in the theory of algebraic invariants and also became his thesis advisor. The thesis was defended in 1885, and Hilbert was appointed to the position of full professor in 1893. He solved Gordan's problem [1] (a problem in the theory of algebraic invariants) around 1888. Hilbert's revolutionary work was generally recognized and accepted afterward. Hilbert saw the number theory as "a building of rare beauty and harmony" [1]. At the German Mathematical Society Annual Meeting in 1893, Hilbert and Minkowski were selected to prepare a report on the current situation in the theory of numbers. It was decided that the report should consist of two separate parts. Minkowski would treat rational number theory, while Hilbert would deal with algebraic number theory. Hilbert subsequently became the world's leading mathematician. His interesting and fruitful activity was his discussion with Henri Poincaré on the reciprocal relationship between analysis and physics. This inspired him to present the list of unsolved problems in his major lecture at the Second International Congress on Mathematics in Paris in 1900. Hilbert wanted to clarify the principles of mathematical reasoning and formalize the axiomatic system that would encompass all mathematics. Hilbert hoped this was the way to obtain the greatest possible objectivity and exactness in mathematics. He says, "There can no longer be any doubt about proofs. The deductive method should be completely clear" [2]. Robert Rosen [3] has a condensed comment about this attitude:

> Hilbert and his formalistic school argued that what we have called *semantic truth could always be effectively replaced by more syntactic rules*. In other words, any external referent, and any quality thereof, could be pulled into a polysynaptic system. By a purely syntactic system, they understood: (1) a finite set of *meaningfulness* symbols, an alphabet; (2) a finite set of rules for combining these symbols into strings or formulas; (3) a finite set of production rules for turning given formulas into new ones. In such a purely syntactic system, *consistency is guaranteed*.

In 1931 Gödel and his incompleteness theorems effectively demolished the formalist program.

Kurt Friedrich Gödel (1906–1978) was born in Brno (in the Czech Republic). In general, Gödel's childhood was a happy one. However, according to his brother, he was "timid and could get upset quickly." At the age of six, Gödel experienced a dose of rheumatic fever. Although he made a full recovery, he believed that he had suffered "permanent heart damage." This was one of the earliest signs of Gödel's later preoccupation with his health. In 1924, he achieved the highest marks in the Deutsches Staats-Realgymnasium. For half a century, he was to be based at the University of Vienna. In 1929 Gödel became an Austrian citizen. Initially, he had been uncertain what to study: physics or mathematics. Because of Gödel's taste for precision, he inclined to mathematics. He was also greatly influenced by the number theorist Philipp Furtwängler. Olga Taussky-Todd, in her reminiscences of Gödel [4], noted that "Gödel hardly ever spoke but was quick to see problems and to point the way to a solution. It was evident that he was exceptionally talented." In her memoirs, Taussky said "that one could talk to Gödel about any problem; he was always clear about the issue and explained matters slowly and calmly." At this time, a mathematician of the new generation, Hans Hahn, became Gödel's principal teacher. Hahn had a number of areas of interest: logic, the foundations of mathematics, and the philosophy of sciences. The Hahn-Banach extension theorem remains well known today. In addition, Hahn was one of the founders of what later became known as *The Vienna Circle* (Der Wiener Kreis). It was probably Hahn who invited Gödel to circle sessions. The circle was engaged in the reading of Wittgenstein's *Tractatus* [5] when Gödel took part in 1926. He attended meetings until 1928. Gödel later had to admit that "from the very beginning of his participation, he was not sympathizing with the circle's views." After 1928 Gödel maintained contacts with a number of members of the Vienna Circle, in particular Rudolph Carnap. He rejected the Viennese empiricism of Hahn and Moritz Schlick, claiming that "that there were truths to be found not just in the empirically perceivable but perhaps beautiful and enduring in the realm of abstract conceptions, where they awaited human discovery not through tangible perception but by thought alone". Gödel's direction of creative work was influenced mainly by Carnap's lectures on Mathematical Logic and the book *Principles of Theoretical Logic* [6] by David Hilbert and Wilhelm Ackermann. At that time, an open problem was whether a system of axioms for the first-order predicate calculus was complete—whether every logically valid statement was a theorem of that calculus. Gödel gave a positive solution to the completeness problem, and this significant achievement marked the beginning of his research career. The research was completed and presented in his doctoral dissertation at the University of Vienna in the summer of 1929. In February 1930, the degree was granted. This year marked Gödel's *annus mirabilis*, when he publicly announced his great result for the first time. It was during "the conference on the foundations of mathematics" in Königsberg in 1930 when Gödel, a relatively unknown graduate student, uttered a few words indicating that he had proof of the incompleteness of arithmetic. He was ignored by almost everyone present; the only one to pay him any attention was the mathematician John von Neumann. It happened at one of the final sessions. The rest is history.

The scientific community accepted his results steadily but slowly. Still, the young Gödel had to spend time in psychiatric hospitals. He had a history of nervous breakdowns caused by hard work. In 1940, Gödel finally arrived in the United States after a journey which took him across Siberia, Japan, and the vast Pacific. He went to Princeton, never returned to Vienna, and remained bitter about his status in Austria from 1939 to 1940. He placed the blame for this situation more on Austrian "schlamperei" than on the cruel Nazi conditions. On his 60th birthday in 1966, he was offered an honorary membership in the Austrian Academy of Sciences, which he declined.

In summary, there were two mathematical systems in the first third of the twentieth century. They were so extensive that it was assumed that every mathematical proposition could be proved or disproved within the system. The great logician Kurt Gödel amazed the entire mathematical world with the paper published in 1931 [7], showing that it was not the case that every mathematical proposition could be either proved or disproved within these systems.

> The development of mathematics in the direction of greater precision has led to large areas of it being formalized, so that proofs can be carried out according to a few mechanical rules. The most comprehensive formal systems to date are, on the one hand, the *Principia Mathematica* of Whitehead and Russell, and, on the other hand, the Zermelo-Fraenkel *System of Axiomatic Set Theory*. Both systems are so extensive that all methods of proof used in mathematics today can be formalized in them, i.e., can be reduced to axioms and rules of inference. It would seem reasonable, therefore, to surmise that these axioms and rules of inference are sufficient to decide all mathematical questions that can be formulated in the systems concerned. In what follows it will be shown that this is not the case, but rather that in both of the cited systems, there exist relatively simple problems of the theory of ordinary whole numbers which cannot be decided on the basis of axioms [7].

Gödel explained that the theorems he would prove did not depend on the nature of the two systems, but rather held for an extensive class of mathematical systems. Gödel's results belong to logic, mathematics, philosophy, and physics. His work has made significant contributions to proof theory, connecting classical, intuitionistic, and modal logic, constructive mathematics, and set theory.

2.2 An Informal Proof of Gödel's Incompleteness Theorems of Formal Arithmetic

The understanding of proof requires certain knowledge of mathematical logic, so we assume that the reader is familiar with basic concepts. Kurt Gödel published his proof of the theorem of incompleteness at the University of Vienna in 1931.

Let K be the first-order predicate theory. What Gödel proved in his Ph.D. thesis is the *theorem of completeness*. A formula A is valid if and only if A is a theorem of K. In what follows, the set of natural numbers is denoted by $N = \{0, 1, \ldots, n, \ldots\}$; thus, we have an algebra $N = (N, +, \cdot, 0, 1)$ that is the *standard model* of the *formal theory of numbers* (*formal arithmetic*) N.

A formal theory of numbers is a first-order theory with equality that has one relation symbol R_1^2, a constant a_1, and function symbols f_1^1, f_1^2, f_1^3. Instead of $R_1^2, a_1, f_1^1(t_1), f_1^2(t_1,t_2), f_1^3(t_1,t_2)$, we write $=, 0, t_1', t_1+t_2, t_1 \cdot t_2$. Proper axioms are (1) $(x = y \wedge y = z) \Rightarrow x = z$, (2) $x = y \Rightarrow x' = y'$, (3) $x' \neq 0 (\neg\, x' = 0)$, (4) $x + 0 = x$,

(5) $x + y' = (x + y)'$, (6) $x \cdot 0 = 0$, (7) $x \cdot y' = x + xy$, (8) $(A(0) \land (\forall x)(A(x) \Rightarrow A(x'))) \Rightarrow (\forall x)A(x)$, where $A(x)$ is a formula of N. As it is already known, (8) is called the *axiom of induction*. The terms $0, 0', 0'', 0''', \ldots$ are called *numerals* and denoted by $0, \lceil 1 \rceil, \lceil 2 \rceil, \lceil 3 \rceil, \ldots$ In general, if n is a nonnegative integer, we will let $\lceil n \rceil$ stand for the corresponding numeral $0''''\ldots'$— i.e., for 0 followed by n strokes. Using the above axioms, one can, for example, prove that in $N \vdash \lceil 2 \rceil + \lceil 2 \rceil = \lceil 4 \rceil$.

A number-theoretic relation $R(x_1, \ldots, x_n)$ is said to be *expressible* in N by a formula $R(x_1, \ldots, x_n)$ of N with n free variables such that, for any natural numbers $k_1, \ldots, k_n$,

1. if $R(k_1, \ldots, k_n)$ is true, then $\vdash_N R(\lceil k_1 \rceil, \ldots, \lceil k_n \rceil)$;
2. if $R(k_1, \ldots, k_n)$ is false, then $\vdash_N \neg R(\lceil k_1 \rceil, \ldots, \lceil k_n \rceil)$.

For a number-theoretic function $f(x_1, \ldots, x_n)$, we say that it is *representable* in N if and only if there is a formula $F(x_1, \ldots, x_n, x_{n+1})$ of N with the free variables $x_1, \ldots, x_n, x_{n+1}$ such that, for any $k_1, \ldots, k_{n+1}$,

1. if $f(k_1, \ldots, k_n) = k_{n+1}$, then $\vdash_N F(\lceil k_1 \rceil, \ldots, \lceil k_n \rceil, \lceil k_{n+1} \rceil) \vdash_N (\exists_1 x_{n+1}) F(\lceil k_1 \rceil, \ldots, \lceil k_n \rceil, x_{n+1})$, where $\exists_1 x_{n+1}$ means "there is only one x_{n+1}";
2. if $R(x_1, \ldots, x_n)$ is a relation, then the *characteristic function of* R, $c_R(x_1, \ldots, x_n)$ is defined as

$$c_R(x_1, \ldots, x_n) = \begin{cases} 0 & \text{if } R(x_1, \ldots, x_n) \text{ is true} \\ 1 & \text{if } R(x_1, \ldots, x_n) \text{ is false} \end{cases}. \quad (2.1)$$

It is easy to see that $R(x_1, \ldots, x_n)$ is expressible in N if and only if $c_R(x_1, \ldots, x_n)$ is representable in N. Namely, if $R(x_1, \ldots, x_n)$ is expressible in N by a formula $F(x_1, \ldots, x_n)$, then it is easy to verify that $c_R(x_1, \ldots, x_n)$ is representable in N by the formula $(F(x_1, \ldots, x_n) \land x_{n+1} = 0) \lor (\neg F(x_1, \ldots, x_n) \land x_{n+1} = \lceil 1 \rceil)$. Conversely, if $c_R(x_1, \ldots, x_n)$ is representable in N by a formula $B(x_1, \ldots, x_n, x_{n+1})$, then $R(x_1, \ldots, x_n)$ is expressible by the formula $B(x_1, \ldots, x_n, 0)$.

Representability of functions and relations in N leads to a class of number-theoretic functions that are of great importance in mathematical logic. The following function mappings $N^k \to N$, for all $k \in N$, are called *initial functions*: (1) the zero function $Z(x) = 0$, for all x; (2) the successor function $x' = x + 1$, for all x; (3) the projection function $I_i^n(x_1, \ldots, x_n) = x_i$, for all $x_1, \ldots, x_n$.

The following are rules for obtaining new functions from given ones:

1. Substitution: $f(x_1, \ldots, x_n) = g(h_1(x_1, \ldots, x_n), \ldots, h_m(x_1, \ldots, x_n))$, f is said to be obtained by *substitution* from the functions $g(y_1, \ldots, y_n)$, $h_1(x_1, \ldots, x_n), \ldots, h_m(x_1, \ldots, x_n)$.

2. Recursion: $f(x_1, \ldots, x_n, 0) = g(x_1, \ldots, x_n); f(x_1, \ldots, x_n, y_{n+1}) = h(x_1, \ldots, x_n, y, f(x_1, \ldots, x_n, y))$. If $n = 0$, we have $f(0) = k$, $f(y + 1) = h(y, f(y))$; f is said to be obtained from g and h by *recursion*.

3. μ-operator: Assume that $g(x_1, \ldots, x_n, y)$ is a function such that for any $x_1, \ldots, x_n$ there is at least one y such that $g(x_1, \ldots, x_n, y) = 0$. We denote the least number y such that $g(x_1, \ldots, x_n, y) = 0$ by $\mu_y(g(x_1, \ldots, x_n, y) = 0)$. Let $f(x_1, \ldots, x_n) = \mu_y(g(x_1, \ldots, x_n, y) = 0)$. Then f is said to be obtained from g by means of the μ-operator if the given assumption about g holds.

A function f is said to be *primitive recursive* if and only if it can be obtained from the initial functions by any finite number of substitutions and recursions. A function f is said to be *recursive* if and only if it can be obtained from the initial functions by any finite number of applications of substitutions, recursions, and μ-operator. Almost all elementary number-theoretic functions we know are primitive recursive. However, there are recursive functions that are not primitive recursive.

A relation $R(x_1, \ldots, x_n) \subseteq N^n$ is *recursive* if $c_R(x_1, \ldots, x_n)$ is a recursive function. It is very important for the proof of Gödel's theorem to have the following assertion: *every recursive relation is expressible in N*. This is a corollary of the theorem: *every recursive function is representable in N*.

Now we are ready to introduce the ingenious idea of Gödel, the so-called *Gödel numbering*. Let T be a first-order theory. With each symbol v of T, we correlate a positive integer $g(v)$ called Gödel number of v in the following way:

$$g(() = 3, g()) = 5, g(,) = 7, g(\neg) = 9, g(\Rightarrow) = 11;$$

$$g(x_k) = 5 + 8k, \text{ for } k = 1, 2, \ldots;$$

$$g(a_k) = 7 + 8k, \text{ for } k = 1, 2, \ldots;$$

$$g(f_k^n) = 9 + 8(2^n 3^k), \text{ for } k, n \geq 1;$$

$$g(A_k^n) = 11 + 8(2^n 3^k), \text{ for } k, n \geq 1. \tag{2.2}$$

Obviously, different symbols have different Gödel numbers, and every Gödel number is an odd positive integer. *Examples.* $g(x_2) = 21$, $g(a_4) = 39$, $g(f_1^2) = 105$, $g(A_2^1) = 155$. If $u_1 u_2 \ldots u_r$ is an expression, we define Gödel number to be $g(u_1 u_2 \ldots u_r) = 2^{g(u_1)} \cdot 3^{g(u_2)} \ldots p_{r-1}^{g(u_r)}$, where p_i is the ith prime and $p_0 = 2$. For example, $g(A_1^2(x_1, x_2)) = 2^{g(A_1^2)} \cdot 3^{g(()} \cdot 5^{g(x_1)} \cdot 7^{g(,)} \cdot 11^{g(x_2)} \cdot 13^{g())} = 2^{107} \cdot 3^3 \cdot 5^{13} \cdot 7^7 \cdot 11^{21} \cdot 13^5$.

Different expressions have different Gödel numbers by the uniqueness of the factorization of integers into primes. If we have an arbitrary finite sequence of expressions $u_1, u_2, \ldots, u_r$, we can assign a Gödel number to this sequence by setting $g(u_1, u_2, \ldots, u_r) = 2^{g(u_1)} \cdot 3^{g(u_2)} \ldots p_{r-1}^{g(u_r)}$.

Let T be any theory with the same symbols as N. Then T is said to be ω-consistent if and only if, for every formula $A(x)$ of T, if $\vdash_T A(\lceil n \rceil)$ for every natural number n, then it is not the case that $\vdash_T (\exists x) \neg A(x)$.

It is easy to see that if T is ω-consistent, then T is consistent. Namely, if T is ω-consistent, consider any formula $A(x)$ that is provable in T, e.g., $x = x \Rightarrow x = x$. In particular, $\vdash \lceil n \rceil = \lceil n \rceil \Rightarrow \lceil n \rceil = \lceil n \rceil$, for all natural numbers n. Therefore, $(\exists x) \neg (x = x \Rightarrow x = x)$ is not provable in T. T is consistent by the tautology $\neg A \Rightarrow (A \Rightarrow B)$ because in case T is inconsistent, every formula B would be provable in T.

Define a relation $R := \text{Proof} \in N \times N$ in the following sense: $\text{Proof}(a, b) \leftrightarrow b$ is the Gödel number of a proof of the formula whose Gödel number is a. $\text{Proof}(x_1, x_2)$ is recursive relation because of the uniqueness of Gödel numbering. Let $Proof(x_1, x_2)$ be a formula that represents the relation $\text{Proof}(x_1, x_2)$ in N. $Proof(x_1, x_2)$ is a formula $A_1^2(x_1, x_2)$ of N, of course. The following formula of N

$$(\forall x_2) \neg Proof(x_1, x_2) \tag{2.3}$$

is crucial in these considerations. Let m be the Gödel number of the formula (2.3), and let

$$(\forall x_2) \neg Proof(\lceil m \rceil, x_2) \tag{2.4}$$

be a formula in the language of N.

Theorem of Incompleteness [7].

(1) If N is consistent, then $\not\vdash (\forall x_2) \neg Proof(\lceil m \rceil, x_2)$.
(2) If N is ω-consistent, then $\not\vdash \neg (\forall x_2) \neg Proof(\lceil m \rceil, x_2)$.

Proof could be demonstrated in the following way:

(1) Suppose N is consistent and $\vdash (\forall x_2) \neg Proof(\lceil m \rceil, x_2)$. Let k be the Gödel number of a proof of this formula. Therefore, $\text{Proof}(m, k)$ is true and thus

$$\vdash Proof(\lceil m \rceil, \lceil k \rceil).$$

However, from the assumption, we have $\vdash (\forall x_2) \neg Proof(\lceil m \rceil, x_2)$. From the logical axioms of N, we have, for $x_2 = \lceil k \rceil$, $\vdash \neg Proof(\lceil m \rceil, \lceil k \rceil)$. N is consistent, thus a contradiction.

(2) Assume that N ω-consistent and that $\vdash \neg\ (\forall\ x_2)\ \neg\ Proof(\ulcorner m\urcorner, x_2)$ i.e., $\vdash(\exists\ x_2)Proof(\ulcorner m\urcorner, x_2)$. N is also consistent, so that not $\vdash(\forall\ x_2)\ \neg\ Proof(\ulcorner m\urcorner, x_2)$. Therefore, for every natural number n, n is not the Gödel number of proof in N of $(\forall\ x_2)\ \neg\ Proof(\ulcorner m\urcorner, x_2)$, i.e., for every n, $Proof(m, n)$ is false. So, for every n, $\vdash \neg\ Proof(\ulcorner m\urcorner, \ulcorner n\urcorner)$. Let $A\ (x_2) := \neg\ Proof(\ulcorner m\urcorner, x_2)$. Then for every n, $\vdash A(\ulcorner n\urcorner)$. By ω-consistency of N, it follows that not $\vdash(\exists\ x_2)\ \neg\ \neg\ Proof(\ulcorner m\urcorner, x_2)$. Hence, not $\vdash(\exists\ x_2)Proof(\ulcorner m\urcorner, x_2)$. But this is a contradiction with $\vdash(\exists\ x_2)Proof(\ulcorner m\urcorner, x_2)$.

As a consequence of the Theorem of Incompleteness, Gödel proved the following:

Theorem of Consistency of N [8]. If N is consistent theory, there is no proof in N of its consistency.

This fact ruined Hilbert's Program. To conclude, if we can prove that $\mathbf{N} = (N, +, \cdot, 0, 1)$ is a model of N, then we have the consistency of N. Therefore, we "believe" that $\mathbf{N}$ is the standard model of N.

2.3 Gödel's Incompleteness Theorems as a Metaphor. Real Possibilities and Misrepresentation in their Applications

Gödel's theory states that mathematics has limitations that do not mean that it is flawed in any way. More generally, it tells us about the limitations of our knowledge or the limitations of broadening or completing our knowledge. Gödel's theorem is an example of vertical progress in mathematics, just as Einstein's theory of relativity and Planck's separation of the discrete from the continuous world are examples in physics (see subchapter 1.2). Their discoveries can be described as an improvement of something that already existed to something new (but surprising). These contributions have in common that they have moved the boundaries of science and mathematics. Metaphorically speaking, as in magical realism novels, suppose the spirit were to appear during one's dinner. Not surprisingly, these circumstances would look unrealistic. However, if one said, "Spirit, this cake is delicious," it would already be magical realism—that is, you would accept the new situation. If one understood and adapted to this newly emerged world, one's perception would be exceptional (i.e., one would have the ability to make vertical progress) (see Pekić's comment in subchapter 1.1). So, there is no mixing of different levels of understanding, just a movement toward a higher level, while the principle of correspondence strongly connects these steps. In other words, in the region as far as physics has come, the question is not whether there are "holes" but where the limitations of knowledge in physics are.

After his discovery in 1931, Gödel was misrepresented by postmodernists, mainly by philosophers who wanted to show that everything was meaningless. Gödel's reasoning and intuition opened the door for "unproved truths," which seemed highly counterintuitive. Mathematicians and philosophers did not initially recognize the profound consequences of Gödel's results. As time passed, experts in other sciences noticed similar situations in their fields, and Gödel's theorem became *a metaphor* for the systems whose consistency could not be proved within themselves. What did Gödel invent in the abstract universe of mathematics? He proved that 1. some matters in that universe could not be proved nor disproved, such as axioms; 2. these matters could not be proved *within* that universe. This is a problem of self-reference (an issue seen in Russell's paradox about sets) expressed through the famous self-referential sentence (the liar paradox): "This sentence is false," creating a logical circularity. If the sentence is true, then it cannot be true; if it is false, it must be true. Gödel applied a similar logic to the whole system of mathematics but with the sentence "This statement is unprovable" and then made a conversion into a number statement about numbers by using a code system later entitled "Gödel numbering." Finally, he found that this proposition could not be proved within a system. In the previous subchapter, we gave informal proof of these theorems. For further reading and a specific look at the insight of Gödel's theorems, we recommend the papers by Hu [9] and Johnstone [10].

Since the lair paradox is a trivial problem, why is it a deep problem? The reason is that its solving is a part of a broad project of understanding truth that involves finding a theory of truth, definition of truth, or proper analysis of the concept of truth [11]. Therefore, although we have no evidence about the direct applications of Gödel's theorems in physics, they are deeply rooted indirectly. Perhaps the following reasoning has the attributes of a paradox. Although Gödel's incompleteness theorems have not yielded results in physics, they are perhaps much more important for physics than mathematics itself. They refer only to a part of mathematics; however, in physics, these theorems, despite their inapplicability, can contribute to defining the boundaries of knowledge. In his famous book *Gödel, Escher, Bach: An Eternal Golden Braid* [12], Douglas Hofstadter illustrates Gödel's world by following his preference for metaphors, analogies, and images (See also [13]).

Let us now return to subchapter 1.1, where we mentioned some examples of how one can question the possible existence of the theory of everything. We also mention Gödel's work in the context of limits supporting "the end of physics" in subchapter 1.2. One should never forget that Gödel's theorem deals with formal systems, and it is questionable whether physical theories are such systems in the mathematical sense. One way to implement Gödel's work is to map the quantum field theory onto number theory [14–15], which is possible under certain assumptions. More often, physicists look for inspiration in Gödel's work and work in *analogy*, although sometimes they do not point out that it is just an analogy. Murphy [16] gives several examples of physicists who work on the "incompleteness" of physics that has nothing to do with Gödel's incompleteness. However, in other cases, they admit that it is just an analogy that might lead to new results [17]. One can say that Gödel's theorems can be treated as a guideline and, as such, are incredibly valuable.

2.4 Gödel's Work in Physical Problems and Computer Science

Physicists, science philosophers, and historians all agree that Einstein's theory of relativity (special and general) is one of the most outstanding achievements of the human mind not only in the twentieth century but of all time (see subchapter 1.3). Kurt Gödel was always interested in physics and established a great relationship with Albert Einstein at Princeton. They had long conversations about relativity theory, so it was no surprise when Gödel offered his contribution to general relativity. Many compliments are given to it because it is mathematically consistent but still leads to many interesting consequences. His contribution to physics includes a scientific paper published in the most valuable journal *Reviews of Modern Physics* and more philosophically inclined work in a collection of papers [18–19].

To explain his contribution and its importance shortly, we must introduce some concepts and accompanying terminology. We use the term *reference frame* for a system of coordinates within which we determine the position of an object. It is usually the standard Cartesian system with coordinates x, y, z, or, more conveniently, x_1, x_2, x_3. This combination defines *an event* if we add time t to the coordinates. The theory of special relativity is based on two postulates that summarize experiment results. One is that the laws of physics appear no different in all reference frames moving uniformly (with constant speed) with respect to each other. The other postulate is that the velocity of light in the vacuum is always the same, independently of the possible motion of the light source or the observer, which is the largest possible velocity in the universe. If one respects these two postulates, one can find the most important rules for transforming physical quantities when the observer moves from one reference frame to another. We usually consider two reference frames with parallel axes and assume that one of the frames (S') moves with respect to the other one (S) with constant (uniform) velocity along one of the axes (x_1). We need to determine how the coordinates and time in the system S' are expressed in terms of the coordinates and time in S. In classical physics, these are so-called Galilean transformations, but they take a different form here: Lorentz transformations. x_1' is expressed in terms of

x_1 and t, but more importantly, we have a new quality: t' that is expressed in terms of both x_1 and t. This implies that time flows differently in a moving frame, which is one of the many fascinating consequences of Lorentz transformations.

One of the essential ideas of relativity is that spatial coordinates and time are entirely equivalent in the description of any object or phenomenon, which means that time should be treated as the fourth coordinate together with three spatial coordinates. (Actually, to have the same dimensionality, the fourth coordinate is ct.) So, we deal with so-called *space-time* instead of just space. It is the three-dimensional Euclidean space with a fourth coordinate added. In such a four-dimensional generalization (the Minkowski space), each point is defined by these four coordinates. The set of these points describing specific motion defines a *worldline*. We choose the set of straight lines for which the distance from the origin equals precisely ct, and this set defines the border of a *light cone*. This is illustrated in Figure 2.1 for a space with two spatial dimensions and time. The worldlines lying inside the cone are entitled time-like, and those outside the cone are called space-like.

Albert Einstein generalized special relativity to the general theory of relativity by including gravity in the following way: he assumed that the presence of masses changes the geometry of space-time since the origins

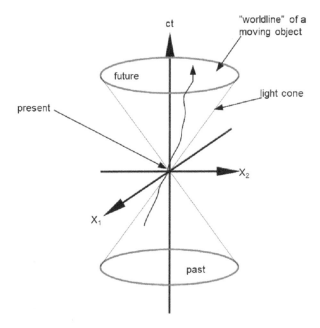

FIGURE 2.1
The Minkowski space with the two spatial and one temporal coordinate interval. (Figure courtesy of Miloš Kapor.)

of gravitational force are masses. Mathematically, he related a *metric tensor* describing the shape of space-time (its curvature) to the tensor describing the distribution of mass and energy in this space using Einstein's field equations. There are few exact solutions of these equations. Among them, some describe the expanding universe that agrees well with present experimental data (i.e., astronomical observations).

Now we come to Gödel's contribution. Gödel used Einstein's equations to propose the rotating universe model for which an exact solution exists. It is an important achievement by itself; however, there is more to it than meets the eye. This solution has one strange property: it allows the existence of closed time-like worldlines. Many physical arguments for the existence of anisotropy in time imply that time flows only in one direction that we choose to call the future (see also subchapter 3.1). It is conveniently named the *arrow of time*. The solutions proposed by Gödel indicate that motion can occur in both directions in time, or better formulated, that there is no privileged direction in time. For the moment, this is treated as speculation since this solution does not lead to the expanding universe. Therefore, this solution and related matters are open for discussion, and more about this theme is mentioned in Chapter 3.

Just as Gödel's work offered a new insight into relativity, so did his remarks about computer science. John von Neumann was seriously ill in the Spring of 1956 (he never recovered and died in 1957), and Gödel tried to distract him by sending him a letter [20] with a question that later became known as the "$P = NP$ problem" in computer science. We consider the problems (questions) with answers of the "yes-no" type. It is favorable that computing time for solving a problem polynomially depends on the algorithm's input. This time is called *polynomial time*. The general class of questions for which an algorithm can provide an answer in polynomial time is denoted by "P" or "class P." Solutions to these problems can be computed in a polynomial amount of time compared with the complexity of the problem. If the algorithm may proceed according to several possibilities at any step of computation, then such an algorithm is called a *nondeterministic algorithm*. The computation of the nondeterministic algorithm is a tree whose branches correspond to different possibilities. The algorithm accepts the input if a particular branch of computation leads to the final state. The class of questions for which the nondeterministic algorithm can provide an answer is denoted by NP. $P = NP$ means that for every problem in NP, there must be an algorithm of polynomial complexity that offers a solution of the problem. Many problems are more complex than the problems of the class NP. One can, by his experience, probability, intuition, or visual thinking, guesses the solution of a highly complex problem. That was the reason why Gödel considered that humans differed from machines. It is challenging to find the solutions of NP problems. The computation may take billions of years, but it is easily checked once the solution is found. The $P = NP$ conjecture is the most critical problem in computer science and one of the most significant ones in mathematics. It is one of the *Millennium Prize Problems* selected by the Clay Mathematical Institute, which offers a $1 million prize for the correct solution of any NP problem. The class NP is rich and includes thousands of NP problems in mathematics and many other sciences. An example of an NP-class problem is the so-called SAT problem.

SAT is the set of satisfiable formulas of sentential logic. To determine whether a formula of n variables is satisfiable involves forming 2^n lines of the formula's truth table. If one microsecond is needed for a line of the truth table of 80 propositional variables, then the execution time of the algorithm is 2^{80} microseconds, which exceeds the universe's age. If $P = NP$, there is a simple algorithm to decide any formula from SAT. The set of Hamiltonian graphs is in NP—that is, we might not know the fast polynomial algorithm, but we do know how to show that the graph has a Hamiltonian cycle. After many decades of the struggle with $P = NP$, we know that it is a tough problem and that we need scientists of Gödel's and von Neumann's caliber to attack this problem seriously. So, Kurt Gödel was far ahead of his time when he posed this problem.

References

[1] C. Reid, *Hilbert*. New York; Berlin, Germany; Vienna, Austria: Springer-Verlag, 1996.

[2] G. J. Chaitin, *Thinking About Gödel and Turing: Essays on Complexity, 1970–2007*. Singapore: World Scientific, 2007.

[3] R. Rosen, *Life Itself: A Comprehensive Inquiry into the Nature, Origin, and Fabrication of Life*. New York, NY, USA: Columbia University Press, 1991.

[4] O. Taussky-Todd, "Remembrances of Kurt Gödel," *Eng. Sci.*, vol. 51, no. 2, pp. 24–28, 1988.

[5] L. Wittgenstein, *Prototractatus: An Early Version of Tractatus Logico-Philosophicus*, B. F. McGuinness, T. Nyberg, and G. H. von Wright, Eds., London, U.K.: Routledge and Kegan Paul, 1971.

[6] D. Hilbert and W. Ackermann, *Grundzüge der theoretischen Logik* (Grundlehren der mathematischen Wissenschaften 27), 4th ed. New York; Berlin, Germany; Vienna, Austria: Springer-Verlag, 1972.

[7] K. Gödel, "On formally undecidable propositions of Principia Mathematica and related systems I," (in German), *Mon. Hefte Math.*, vol. 38, no. 1, pp. 173–198, Dec. 1931, doi:10.1007/BF01700692.

[8] K. Gödel, "Discussion on the foundation of mathematics," (in German), *Erkenntnis*, vol. 2, pp. 147–151, Dec. 1931, doi:10.1007/BF02028146.

[9] J. Hu, "Gödel's proof and the liar paradox," *Notre Dame J. Form. Log.*, vol. 20, no. 3, Jul. 1979.

[10] A. Johnstone, "Self-reference and Gödel's theorem: A Husserlian analysis," *Husserl Studies*, vol. 19, no. 2, pp. 131–151, Jun. 2003, doi:10.1023/A:1024843725932.
[11] M. Glanzberg, "Truth," in *The Stanford Encyclopedia of Philosophy*, E. N. Zalta, Ed., Stanford University Press, 1997–. Available: https://plato.stanford.edu/archives/sum2021/entries/truth/
[12] D. R. Hofstadter, *Gödel, Escher, Bach: An Eternal Golden Braid*. Brighton, UK: Harvester Press, 1979.
[13] D. R. Hofstadter, "Analogies and metaphors to explain Gödel's theorem," *Two-Year Coll. Math. J.*, vol. 13, no. 2, pp. 98–114, Mar. 1982, doi:10.2307/3026485.
[14] M. Pitkänen, "Physics as generalized number theory: Infinite primes," unpublished.
[15] I. V. Volovich, "Number theory as the ultimate physical theory," *P-Adic Num. Ultrametr. Anal. Appl.*, vol. 2, no. 1, pp. 77–87, Jan. 2010, doi:10.1134/S2070046610010061.
[16] P. A. Murphy. "Do Gödel's theorems impact negatively on physics?." *Cantorsparadise.com*. https://www.cantorsparadise.com/do-g%C3%B6dels-theorems-impact-negatively-on-physics-ed09734b424c (accessed Mar. 24, 2022).
[17] J. M. Myers and F. H. Madjid, "Incompleteness theorem for physics," 2019, *arXiv*:1803.10589.
[18] K. Gödel, "An example of a new type of cosmological solutions of Einstein's field equations of gravitation," *Rev. Mod. Phys.*, vol. 21, no. 3, pp. 447–450, Jul. 1949, doi:10.1103/RevModPhys.21.447.
[19] K. Gödel, "A remark about the relationship between relativity theory and idealistic philosophy," in *Albert Einstein: Philosopher-Scientist* (The Library of Living Philosophers 7), P. A. Schilpp, Ed., Evanston, IL, USA: Library of Living Philosophers, 1949, pp. 557–562.
[20] S. Feferman, J. W. Dawson, W. Goldfarb, C. Parsons, and W. Sieg, Eds., *Kurt Gödel: Collected Works*, vol. 5, London, U.K.: Oxford University Press, 2013.

3

Time in Physics

3.1 Time in Philosophy and Physics: Beyond Gödel's Time

Time is one of the most enigmatic phenomena we experience and try to explain throughout our lives. It is a concept, and only philosophy deals with pure concepts, which is its exclusivity. Although time is perhaps the most complex philosophical problem, Rüdiger Safranski, a German philosopher, presents the understanding of time and its history simply [1], offering several perspectives of time: (1) psychological as the most attractive but also most superficial one; (2) literary as the closest to philosophy; and (3) natural-scientific as "too short." Many aspects of the above perspectives (time in philosophy, time in physics, functional time, endo-time, etc.) are discussed in [2]. Isaac Newton introduced what provided the theoretical basis for Newtonian mechanics: the concept of absolute time and space: "Absolute, true and mathematical time, of itself, and from its own nature flows evenly regardless of anything external, remains always similar and immovable" [3]. We cite an interpretive text about Kant's views of time [4] offered by Dorato et al. [5]: "Kant's views of time, in such a way as to distillate three conditions that are together necessary for the ideality of time in Kant's sense: (i) time must be no substantial, and the resulting relationism must be constructed in such a way that both (ii) the difference between past and future and that (iii) between earlier and later than, must be mind-dependent." Schopenhauer was so fascinated by Kant's understanding of time that he noticed that we had "carried" our heads in time and space before Kant, while we "carried" time and space in our heads after Kant. It is also inevitable to mention McTaggart's work "The Unreality of Time" [6], published in 1908. He is famous for arguing that time is unreal because our descriptions of time are either contradictory, circular, or insufficient. Einstein's interpretation of Lorentz's transformation introduced a concept different from absolute time: it was that time depended on a reference frame.

We use the following points to introduce an important concept about time metaphorically. Because of the reliance on hypotheses, induction, and their limitations, physicists always have the problem of 1. choosing the right direction in research; and 2. "dealing" with time. This situation can be visualized on the bifurcation map of the differential logistic equation (the logistic equation is chosen since it appears to be a master equation in physics and science [2]). The map consists of black and white spots corresponding to chaos and stability. Physicists are placed within one of the white spots while trying to move to other white spots, not in an arbitrary direction but in the direction that must be defined by the universe, its content, and the way it evolved. This universal direction was called the arrow of time (i.e., the "one-way direction" or "asymmetry" of time).

> Let us draw an arrow arbitrarily. If as we follow the arrow we find more and more of the random element in the state of the world, then the arrow is pointing towards the future; if the random element decreases the arrow points towards the past. That is the only distinction known to physics. This follows at once if our fundamental contention is admitted that the introduction of randomness is the only thing which cannot be undone. I shall use the phrase 'time's arrow' to express this one-way property of time which has no analogue in space.
>
> [7]

1. *The thermodynamic arrow of time* is related to the entropy of an isolated system that either remains constant (in rare cases) or increases with time (Figure 3.1). Because of the connection between its direction and entropy, the past (lower entropy) and the future (higher entropy) can be distinguished.

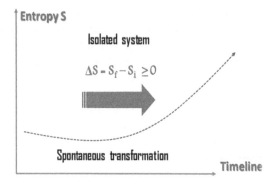

FIGURE 3.1
The representation of entropy as an arrow of time. (Reproduced by permission from [8].)

2. Starting from the fact that the arrow of time is defined by the direction of time in which entropy increases, Hawking [9] shows that the direction of the time of the universe's expansion determines *the cosmological arrow of time*. 3. The "collapse of the wave function," which, according to present interpretations [10], occurs in the process of measurement, is definitely irreversible because, due to the collapse, the complete information from the wave function is lost without any possibility of being reconstructed. In this regard, the process is time-irreversible, and *the quantum arrow of time* is formed. In quantum physics, past measurements partially constrain future measurements. Accordingly, the notion of a quantum state is unaffectedly "time-asymmetric." Asymmetry is separated by the "symmetrization of time" in [11] to avoid this concept. The authors introduced the two-state vector formalism of quantum mechanics and later extended their work [12–13]. 4. In theory, equations allow waves to be convergent, while this situation has never been observed in nature. This radiation flow discrepancy is regarded as *the radiative arrow of time*. 5. Despite the CPT theorem, physical laws are not invariant to CP operations in specific subtle interactions [14], which characterizes *the weak arrow of time*. 6. *The psychological arrow of time* was first mentioned by Stephen Hawking in his book *A Brief History of Time* [15] and later in the series of his lectures [16–17]. "Disorder increases with time because we measure time in the direction in which disorder increases" [15] implies that the psychological arrow of time is determined by the thermodynamic one. Therefore, we can define the direction of the psychological arrow of time as our perception of the past and the future: we remember the past (decreased entropy) but not the future (increased entropy). This arrow should align with the well-defined thermodynamic arrow of time [18].

The model of the universe known as Gödel's space-time [19] (see subchapter 2.4) is based on an essential rotation of matter. However, the solution of its equations does not exhibit expansion like the actual universe. Nevertheless, models with both rotation and expansion are also theoretically possible. Precise measurements recently indicated that the actual universe does not rotate [20]. This fact neither diminishes nor overshadows questions posed in the context of Gödel's space-time. Gödel's model of the universe, although not the actual model of the universe, is theoretically possible. It allows closed time-like worldlines without a preferred direction of time (see subchapter 2.4). Only a small number of philosophers and people from the scientific community accepted Gödel's view about time based on that solution. In another paper [21], he concludes that time conjectured by philosophers such as Immanuel Kant is ideal. Richmond [22] interprets this inference as the unreal time that does not have objective existence.

Gödel was deeply aware of the ontological nature of time. Then where is Gödel's view about time positioned in philosophy and physics? When we consider Gödel's time, we look at it from the perspective of science or philosophy. In philosophy, the main alternative to presentism, which supports that nothing but the present is real, is eternalism, which supports that any event in the past, the present, and the future exists. The most intriguing point is that Gödel's model allows traveling back in time without an exceeding of the local speed of light. While time in physics is defined as irreversible, Gödel's time is cyclic: objects can return at a certain point in the past. To summarize Gödel's time, we use Riggs's response [22] to Gödel's solution of Einstein's field equations [23]: 1. Gödel's space-time shows that the order of events in time is not always globally consistent in space-times that consist of closed time-like worldlines. 2. Relying on Gödel's model [19], some researchers proposed that time travel was possible without the rotating universe [24]. 3. Our understanding of the laws of physics does not eliminate the possibility of time travel [25]. 4. Since Gödel's space-time results from Einstein's field equations, it is a model of a possible universe [22]. 5. Time travel will be an open question and be excluded from physics.

Like all open questions in physics, the question of backward time travel may be closed in a manner unknown to us at this level of our knowledge. In connection with this dilemma, Borislav Pekić's reasoning [26] about an open question is also interesting: "The existence of spirits in principle does not contradict any law of physics. It is in contrast to the mind that civilization has modeled on empirical evidence for centuries" (see subchapter 1.1).

3.2 Does the Quantum of Time Exist?

We dedicated a complete chapter, a lot of additional material in our previous book [2], and subchapter 3.1 to the concept of time. Accordingly, in this subchapter some other aspects of time in physics are given. In quantum mechanics and classical physics, time is treated as a parameter, rather than a dynamical variable. The simplest explanation of why it is not a dynamical variable is that the action of force in classical physics, special relativity, and quantum mechanics does not influence that time. Thus, there remains the problem of whether time can be quantized. A popular idea of quantization is that "quantum" is the smallest value of a certain quantity ("quantum"), and all other values are multiples of that quantum. (This idea is attributed to Max Planck, who introduced the quantum of the energy of the linear harmonic oscillator to explain the law of the black body radiation. Previously, he used a classical equivalence of the electromagnetic field in a closed volume of space with the system of independent harmonic oscillators.) This is also valid for light ("photons") and Millikan's discovery of charge quantization [27].

The obvious question is whether such a quantity of time exists. There does indeed exist one called the Planck time [28]. First, we introduce the Planck length

$$l_P = \sqrt{\frac{\hbar G}{c^3}} = 1.616 \times 10^{-35} \text{ m} \quad (3.2.1)$$

which is derived in dimensional analysis from three fundamental physical constants: reduced Planck's constant $\hbar$, gravitational constant G, and velocity of light in vacuum c. Is this the smallest possible length? The estimated size of a quark is ~10^{-18} m; thus, the Planck length is 18 orders of magnitude smaller than the smallest particle in the universe (it is supposed that quarks are composed of smaller particles—preons). Although dimensional analysis is a powerful tool in many situations, it is not a reliable method that can provide us with consistent information in the way that the Planck scale can intrinsically contribute to quantum gravity research with strong relevance [29]. Is this precisely the case with the Planck length? This question is quite natural if one considers the ultimate limits of measurement at this scale [30]. In that case, a refined length requires a large momentum according to Heisenberg's uncertainty principle; thus, its gravitational effect becomes strong. Mead [31] finds that the position of a particle cannot be measured with an error less than $\Delta x = \sqrt{G} \, 1.6 \times 10^{-35}$ m while analyzing the effect of gravitation on hypothetical experiments. He suggests that it may be impossible to formulate a fundamental length theory that does not include the effect of gravitation in a significant manner. Alden Mead said that Henry Primakoff, David Bohm, and Roger Penrose supported him in the idea that l_P was a fundamental length during his suffering of the referee trouble (1959–1964) [32]. Frank Wilczek later showed that simple dimensional analysis, combined with a few elementary facts, could lead to profound conclusions [30, 33].

This length has been interpreted in various ways. The simplest description is that it is the limit where known physical laws may fail since it is impossible to go beneath the Planck scale in terms of time or distance, and a new theory combining quantum mechanics and gravity (still unknown) should work. Then, the Planck time t_P

$$t_P = \frac{l_P}{c} = 5.89 \times 10^{-44} \text{ s} \quad (3.2.2)$$

is the time taken by light to cross that distance. According to current concepts, this should be the smallest possible time interval between two related events. The Planck time is generally viewed as an elementary "pixel" of time within which the physics of four-dimensional space-time decomposes into a far larger number of dimensions assumed by superstring theory. Our understanding of space-time becomes ambiguous beyond this time scale. However, it is not the quantum of time because there is no proof that time is a multiple of the Planck time.

One must remember that the concept of quantization can be introduced more strictly: the phenomenon that certain physical quantities have only certain values (the energy of an electron in the hydrogen atom). In addition, the spatial quantization of the directions of a particle's angular momentum and spin in the magnetic field should be kept in mind. All these appear in nature, and the numerous studies of the spectra of noninteracting atoms and molecules prove their existence. We do not know the origin, but we explain their existence by studying the eigenvalue problem of the quantum mechanical operator representing a given physical quantity. Then quantization follows from the requirement that solutions must satisfy certain conditions (mathematical and physical). From this point of view, one needs the quantum of time that, for the moment, should appear only in the theory combining quantum mechanics with gravity [34] (there was a popular such theory [35] that introduced the quantum of time, the so-called chronon).

3.3 Continuous and Discrete Time

The purpose of this subchapter is to summarize how time is treated in dealing with dynamical systems in physics and some related problems. The two recognized perspectives of handling time in dynamical systems, finite and infinitesimal, are formalized throughout the two distinct concepts of time—*discrete* and *continuous*. When it comes to discrete time, the time axis is separated into smaller time segments of fixed length, and the resulting number of points and intervals is finite. Quite the opposite, continuous time involves infinitesimally small time intervals. The changes of a state are modeled by the first derivative of the state function, while higher-order derivatives are used to formalize higher-order changes. Thus, the changes of state variables over time are related to intervals and specified for specific points or to each point of time. An interesting fact is that continuous time can be treated as an imaginary quantity (*imaginary time*). This time is not imaginary in a literal sense; it is just real time that undergoes the Wick rotation. In the theory of many-body systems, the time dependence of physical quantities in Matsubara Green's functions is described by imaginary time. This allows simpler calculations, and Green's functions are related to the standard ones afterward to evaluate the relevant physical properties of a system by using well-elaborated methods (see [36]). This time is a mathematical representation of time in special relativity and quantum mechanics. It may also appear in quantum statistical mechanics and specific cosmological theories.

Physicists need to decide on time treatment because conceptualizing time in dynamical models is not straightforward. Whether we should use continuous or discrete time includes dilemmas that still need to be resolved (for further reading see [2]). Many symmetries in continuous models are lost after discretization. While continuum models are more comfortable, small discrete systems are considerably simpler to manage. Whether nature is essentially discrete or not, most physical models are continuous; therefore, ordinary or partial differential equations are *substituted* with appropriate difference equations. It is more accepted to use discrete difference equation(s) for the model building to avoid the process of 1. obtaining a differential equation that approximates a discrete situation; and 2. approximating that differential equation with a difference scheme (for numerical computing purposes) [37–38]. If some phenomena are already described by equation(s), the corresponding laws can be deduced from symmetry conditions. Since this is not usually the case, we often determine equations from some hypotheses and experimental data [39]. Even if we know which equations should be used, we can still be uncertain about the values of parameters because of the numerous simplifications we designed. The mathematical solution of the corresponding differential or difference equations system is not always physically plausible. One reason is that mathematicians and physicists have different understandings of the term random. The second one is that the solutions may be mathematically correct but physically meaningless [2]. For example, under some conditions that occur in the atmosphere, the energy balance equation can be written in the form of

$$X_{n+1} = A_n X_n - B_n X_n^2, \tag{3.3.1}$$

where X_n is the dimensionless environmental interface temperature, while coefficients A_n and B_n are dimensionless parameters. The solution of (3.3.1) for the specific choice of A_n and B_n is depicted in the upper panel of Figure 3.3.2 (chaotic behavior). According to Kreinovic's opinion [39], the question is whether we can find domain(s) for this equation (the lower panel of Figure 3.3.2) where physically meaningful solutions exist.

The choice of the time step Δt is subject to constraints because Δt must satisfy some physical and numerical conditions. Courant–Friedrichs–Lewy's (CFL) condition [40] must be met to solve certain partial differential equations numerically since it is a necessary condition for convergence. According to this criterion, a simulation gives incorrect results if the time step is not less than a certain time period in many explicit time-marching computer simulations. The choice of the time step depends on how we decide to discretize the system of equations, but spatial and temporal resolutions are connected through the CFL condition.

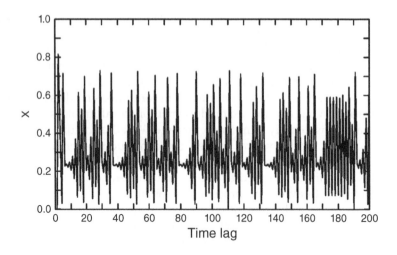

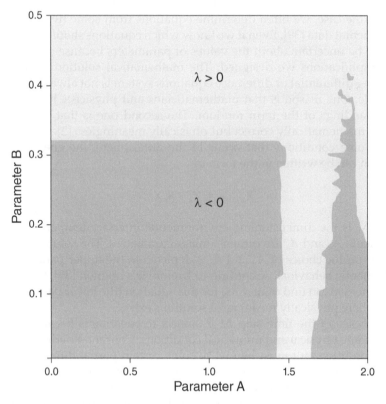

FIGURE 3.2
Upper, the chaotic fluctuations of the dimensionless environmental interface temperature X in (3.3.1). *Lower*, the regions of stable ($\lambda < 0$) and unstable ($\lambda > 0$) solutions of (3.3.1) formed by the Lyapunov exponent's values λ (see the next chapter) and the coefficients $A \in (0, 2)$ and $B \in (0, 0.5)$. (Reproduced by permission from [2].)

Surfaces often have such thermal characteristics that the coefficients of the Equation (3.3.1) may change significantly during the calculation of the dimensionless environmental interface temperature from the energy balance equation [2]. If certain conditions are met, this equation can be transformed into the following logistic equation

$$X_{n+1} = rX_n(1-X_n),$$

$$r = 1+\tau,$$

$$\tau = \frac{\Delta t}{\Delta t_p}, \qquad (3.3.2)$$

where X_n is the dimensionless environmental temperature, $0 < r < 4$, τ is the dimensionless time, Δt is the time step, and Δt_p is the *scaling time* of energy exchange at the environmental interface. The scaling time can take the form of $\Delta t_p = C_g/\Sigma$, where C_g is the heat capacity of the environmental interface per unit area, while Σ is the net energy amount at the environmental interface. We can use Δt_p defined in this manner to choose the time step Δt to solve the energy balance equation numerically (i.e., $\Delta t < 2.57\Delta t_p$) [2]. Using this inequality, we avoid the scenario where a tremendous amount of energy abruptly entering a system cannot be accepted by the environmental interface. This criterion depends only on the net energy of the environmental interface and its thermal properties. In the inequality $\Delta t < 2.57\Delta t_p$, the scaling time Δt_p is the *relaxation time*. The energy exchange at a smaller scale can be noticed when $\Delta t < \Delta t_p$ (i.e., we can avoid chaos).

3.4 Time in Complex Systems

During the preparation of this book, the Nobel Prize in Physics 2021 was awarded to three scientists "for groundbreaking contributions to our understanding of complex systems" [41]. Syukuro Manabe and Klaus Hasselmann were awarded "for the physical modelling of Earth's climate, quantifying variability and reliably predicting global warming" [41]. Giorgio Parisi was awarded "for the discovery of the interplay of disorder and fluctuations in physical systems from atomic to planetary scales" [41]. These were the first occasions on which the Nobel Prize in Physics was awarded to scientists for their research on complex systems, including the climate. In [42], Garisto says, "Unfortunately, by grouping seemingly unrelated research under the vague umbrella of complex systems, the Nobel Committee for Physics puzzled many observers and led to hazy headlines such as 'Physics Nobel Rewards Work on Climate Change, Other Forces.' What links these very different discoveries is, at first, far from clear. But a close examination reveals some connections—about the aims of science and how scientists can tackle seemingly intractable problems." Reactions to the decision of the Nobel Committee were different, which could have been expected. Among them, one reaction was interesting. "There is no clear definition of complex systems," says Kunihiko Kaneko, a physicist at the University of Tokyo. "But roughly speaking, there are many interacting elements, and they often show chaotic or dynamic behavior" [42]. This view is at least encouraging to think about. The Nobel Committee for Physics explained what had motivated them to make this decision and award the scientists who examined "the phenomena we observe in nature emerge from an underlying disorder, and that embracing the noise and uncertainty is an essential step on the road towards predictability" [42]. Suppose we accept that chaos and disorder are synonyms and recognize philosophical elements that we use to describe connections in this text. In that case, we can more clearly understand Saramago's credo in the novel *The Doubled* [43]: "The chaos is order yet to be puzzled out." Our view on the issues mentioned above and comments can be found in subchapters 1.2 and 1.3. It should be noted that Ilya Prigogine was the first to give prominence to complex systems and new structures that emerge owing to the internal self-reorganization in his pioneering papers. He was awarded the Nobel Prize in Chemistry in 1977 "for his contributions to non-equilibrium thermodynamics, particularly the theory of dissipative structures" [44]. Even though chemistry was relatively closed down at the time, chemists noticed the profound nature of his ideas much earlier than physicists.

The above-condensed text about those people who have built the foundations of complex systems is a short introduction to the *time of complex systems* (or how it is briefly and not quite correctly called *complex time*). It has a fundamental meaning because many questions about the temporal asymptotic behavior of complex systems in infinity cannot be answered by performing finite computations. The broad consensus from numerous sources is that time in complex systems can be described as follows: "Time in complex systems operates concurrently at different scales, runs at multiple rates, and integrates the function of numerous connected systems" [45]. This definition is a framework that includes components extending from evident to hidden but important ones. Its potential drawbacks are discussed in papers about complex systems whose number is not negligible. We extend this definition by adding our own experience regarding this quantity at a complex system's spatial scales mentioned in the conclusion of the Nobel Committee for Physics in 2021.

The simulation of complex systems is an important branch that allows us to simulate a complex process and integrate a set of underlying evolution equations. It is difficult for scientists to create an optimal model because interactions between components in a complex system change over time. This becomes even more complicated when the underlying process is unknown. We explain three contexts that reflect the uses of time in complex systems modeling: 1. *Time stepping*. In climate models, the partial differential equations that describe the physical phenomena in the complex earth–atmosphere interaction are discretized numerically in space and time to obtain a solution. Therefore, the emerging time integration strategy (time stepping with a *time step* Δt) is essential for model building [46]. This procedure includes a. different numerical algorithms for the adiabatic fluid dynamics and sub-grid scale diabatic physics; and b. the numerical stability limits of each atmospheric process. The time integration in climate models is performed for each Δt, including smaller time steps Δt_i that satisfy the conditions of physical and numerical stability for different model modules $i = 1, 2, ...N$. To illustrate this point, in the CAM5 climate model convergence tests [47] for $\Delta t = 1800$ s and $N = 9$, the ratios of time sizes to Δt of radiation, deep convection, shallow convection, stratiform cloud macrophysics, stratiform cloud microphysics, vertical remapping, adiabatic fluid dynamics, resolved-scale tracer transport, and explicit numerical diffusion were 2, 1, 1, 1, 1, 1/2, 1/6, 1/6, and 1/18, respectively. 2. *Time-scale separation*. The state of a real system is often determined by more than a few processes that may operate at different time scales. The difficult task is to recognize the sizes of these time scales, at least for the main processes. An additional complication is that interactions change over time—an issue rarely considered in the modeling. An illustrative example is when preys change their behavior to avoid predators at a small or large time scale. One modeling strategy is to introduce a simplification by assuming that all interactions are constant (this can also lead to false estimates). The mathematical background for time-scale separation is Tychonoff's theorem [48], stating that if fast and slow components become more separated when a small parameter approaches zero, the arbitrary product of compact spaces is compact [49]. Fast and slow components categorize processes and reactions at unequal time scales. The reduction of the number of variables and parameters in the model by applying the quasi-steady-state approximation disables the modeling of fast elementary reactions explicitly (note that Michaelis and Menten [50] introduced this idea). Their influence is expressed through the parametrization by nonelementary reaction-rate functions, which is similar to the parametrization in climate models. Because of complex interactions between fast and slow components, this time-scale separation cannot be used for modeling biological systems [51]. Figure 3.3 visualizes particle diffusion paired up with the competitive birth-death interaction as an example of time-scale separation.

3. *Time in functional systems* can be treated in another way. We can change our perspective of time and leave the position of external objective observers by looking at the process itself [52–53]. Because functional systems are regarded as languages that describe how a complex system functions, we

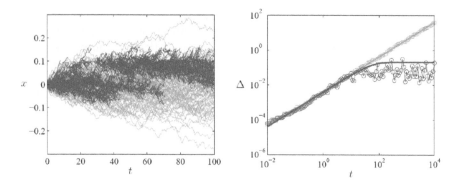

FIGURE 3.3
The simulation of competition-limited diffusion (*dark gray*) contrasted with a collection of N-independent Brownian particles (*light gray*); t is the time. *Left*, the particle trajectories (x); *right*, the mean square distance between pairs of particles (Δ). (Reproduced by permission from [51].)

categorize them as complex systems. In such systems, the time structure differs from the o'clock time and can be considered as *functional time*. Functional time is derived from the actual system and ultimately depends on the quality variation within the system and its development process. The variation of the system's stability influenced by these changes is considered either through its structural stability or synchronization established by measurement or through the methods of nonlinear dynamics. As an illustration, we show the simulation of the functional time formation in biochemical substance exchange between cells (inter-ring arrangement). This time formation is modeled by the system of coupled difference equations [2, 53], and its model has the following features: a. The dependence of the signal strength on distances between cells is computed in the π-cell coordinate system. b. Only in the case when the largest Lyapunov exponent of the system takes negative values the stability of the system is globally synchronized (coded 1, 0 otherwise). c. Signals from the system to an observer's location indicate that the system does not lose its functionality; that functionality can be terminated if so-called cardinal components lose their functionality. If the system does not send any signal from non-cardinal components, it means that it sends signals but not the signal from that component or that a non-cardinal component is on standby for synchronization with other components to send the signal. It is possible to establish functional time in biochemical substance exchange in a multicellular system because the state of a complex system is binary (i.e., either unsynchronized (0) or synchronized (1)). This code is called a *functional time barcode* [2] and shows the "history" of the states of the system when the system evolves in time (Figure 3.4).

Time in Physics 45

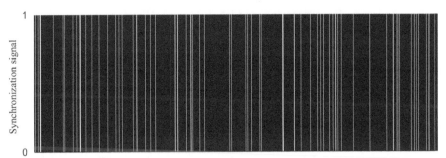

Functional time of biochemical substance exchange in the system of three cells (10,000 iterations)

FIGURE 3.4
The functional time barcode of biochemical substance exchange inter-ring arrangement of cells system. (Reproduced by permission from [53].)

References

[1] R. Safranski, *Zeit: Was sie mit uns macht und was wir aus ihr machen*. Munich, Germany: Carl Hanser Verlag, 2015.

[2] D. T. Mihailović, I. Balaž, and D. Kapor, *Time and Methods in Environmental Interfaces Modelling: Personal Insights* (Developments in Environmental Modelling 29). Amsterdam, The Netherlands; New York, NY, USA: Elsevier, 2016.

[3] I. Newton, *The Principia: Mathematical Principles of Natural Philosophy*, I. B. Cohen and A. Whitman, Trans., Berkeley, CA, USA: University of California Press, 1999.

[4] I. Kant, *Critique of Pure Reason* (The Cambridge Edition of the Works of Immanuel Kant), P. Guyer and A. W. Wood, Eds., Cambridge, U.K.: Cambridge University Press, 1998.

[5] M. Dorato, "Kant, Gödel and relativity," in *Proc. 11th Int. Congr. Log. Methodol. Philos. Sci.*, P. Gardenfors, K. Kijania-Placek, and J. Wolenski, Eds. 2002, pp. 329–346.

[6] J. E. McTaggart, "The unreality of time," *Mind*, vol. 17, no. 4, pp. 457–474, Jan. 1908, doi: 10.1093/mind/xvii.4.457.

[7] A. S. Eddington, *The Nature of the Physical World: Gifford Lectures 1927*. Cambridge, U.K.: Cambridge University Press, 2012.

[8] D. Soubane, M. Garah, M. Bouhassoune, A. Tirbiyine, A. Ramzi, and S. Laasri, "Hidden information, energy dispersion and disorder: Does entropy really measure disorder?," *World J. Condens. Matter Phys.*, vol. 8, no. 4, Nov. 2018, doi: 10.4236/wjcmp.2018.84014.

[9] S. W. Hawking, "Arrow of time in cosmology," *Phys. Rev. D*, vol. 32, no. 10, pp. 2489–2495, Nov. 1985, doi: 10.1103/PhysRevD.32.2489.

[10] C. Cohen-Tannoudji, B. Diu, and F. Laloë, *Quantum Mechanics*, 2nd ed. Weinheim, Germany: Wiley-VCH, 2019.

[11] Y. Aharonov, P. G. Bergmann, and J. L. Lebowitz, "Time symmetry in the quantum process of measurement," *Phys. Rev.*, vol. 134, no. 6B, pp. B1410–B1416, Jun. 1964, doi: 10.1103/PhysRev.134.B1410.

[12] Y. Aharonov and L. Vaidman, "The two-state vector formalism: An updated review," in *Time in Quantum Mechanics* (Lecture Notes in Physics 734), J. G. Muga, R. Sala Mayato, and Í. L. Egusquiza, Eds., 2nd ed. New York; Berlin, Germany; Vienna, Austria: Springer-Verlag, 2008, pp. 399–447.

[13] Y. Aharonov, E. Cohen, and T. Landsberger, "The two-time interpretation and macroscopic time-reversibility," *Entropy*, vol. 19, no. 3, p. 111, Mar. 2017, doi: 10.3390/e19030111.

[14] J. H. Christenson, J. W. Cronin, V. L. Fitch, and R. Turlay, "Evidence for the 2π decay of the K_2^0 meson," *Phys. Rev. Lett.*, vol. 13, no. 4, pp. 138–140, Jul. 1964, doi: 10.1103/PhysRevLett.13.138.

[15] S. W. Hawking, *A Brief History of Time: From the Big Bang to Black Holes*. New York, NY, USA: Bantam Dell, 1988.

[16] S. W. Hawking, "The no-boundary proposal and the arrow of time," in *Physical Origins of Time Asymmetry*, J. J. Halliwell, J. Pérez-Mercader, and W. H. Zurek, Eds., Cambridge, UK: Cambridge University Press, 1996, pp. 346–357.

[17] S. W. Hawking, "Gödel and the end of physics" (lecture, Texas A&M University, College Station, TX, Mar. 8, 2002). Available: http://yclept.ucdavis.edu/course/215c.S17/TEX/GodelAndEndOfPhysics.pdf

[18] L. Mlodinow and T. A. Brun, "Relation between the psychological and thermodynamic arrows of time," *Phys. Rev. E*, vol. 89, no. 5, May 2014, Art. no. 052102, doi: 10.1103/PhysRevE.89.052102.

[19] K. Gödel, "An example of a new type of cosmological solutions of Einstein's field equations of gravitation," *Rev. Mod. Phys.*, vol. 21, no. 3, pp. 447–450, Jul. 1949, doi: 10.1103/RevModPhys.21.447.

[20] D. Saadeh, S. M. Feeney, A. Pontzen, H. V. Peiris, and J. D. McEwen, "How isotropic is the universe?," *Phys. Rev. Lett.*, vol. 117, no. 13, Sep. 2016, doi: 10.1103/PhysRevLett.117.131302.

[21] K. Gödel, "A remark about the relationship between relativity theory and idealistic philosophy," in *Albert Einstein: Philosopher-Scientist*, P. A. Schilpp, Ed., Evanston, IL, USA: Library of Living Philosophers, 1949, pp. 557–562.

[22] P. Riggs, "Time travelers," *Inference*, vol. 4, no. 1, May 2018. [Online]. Available: https://inference-review.com/article/time-travelers

[23] A. Richmond, "Time travel, hyperspace and Cheshire Cats," *Synthese*, vol. 195, no. 11, pp. 5037–5058, Nov. 2018, doi: 10.1007/s11229-017-1448-2.

[24] J. R. Gott, *Time Travel in Einstein's Universe: The Physical Possibilities of Travel through Time*. Boston, MA, USA: Houghton Mifflin, 2001, p. 291.

[25] J. Earman, C. Smeenk, and C. Wüthrich, "Do the laws of physics forbid the operation of time machines?," *Synthese*, vol. 169, no. 1, pp. 91–124, Jul. 2009, doi: 10.1007/s11229-008-9338-2.

[26] B. Pekić, *Atlantida*. Belgrade, Serbia: Laguna, 2015.

[27] K. Simonyi, *A Cultural History of Physics*, D. Kramer, Trans., Boca Raton, FL: CRC Press, 2012.

[28] S. Weinstein and D. Rickles, "Quantum gravity," in *The Stanford Encyclopedia of Philosophy*, E. N. Zalta, Ed., Stanford University Press, 1997–. Available: https://plato.stanford.edu/archives/fall2021/entries/quantum-gravity/

[29] D. Meschini, "Planck-scale physics: Facts and beliefs," *Found. Sci.*, vol. 12, no. 4, pp. 277–294, Dec. 2007, doi: 10.1007/s10699-006-9102-3.
[30] F. Wilczek, "Scaling mount Planck I: A view from the bottom," *Phys. Today*, vol. 54, no. 6, pp. 12–13, Jun. 2001, doi: 10.1063/1.1387576.
[31] C. A. Mead, "Possible connection between gravitation and fundamental length," *Phys. Rev.*, vol. 135, no. 3B, pp. B849–B862, Aug. 1964, doi: 10.1103/PhysRev.135.B849.
[32] C. A. Mead, "Walking the Planck length through history," *Phys. Today*, vol. 54, no. 11, p. 15, Nov. 2001, doi: 10.1063/1.1428424.
[33] F. Wilczek, "Walking the Planck length through history," *Phys. Today*, vol. 54, no. 11, p. 15, Nov. 2001, doi: 10.1063/1.4796228.
[34] H.-D. Zeh, *The Physical Basis of the Direction of Time*, 2nd ed. New York; Berlin, Germany; Vienna, Austria: Springer-Verlag, 1992.
[35] P. Caldirola, "The introduction of the chronon in the electron theory and a charged-Lepton mass formula," *Lett. Nuovo Cimento*, vol. 27, no. 8, pp. 225–228, Feb. 1980, doi: 10.1007/BF02750348.
[36] G. D. Mahan, *Many-Particle Physics* (Physics of Solids and Liquids), 3rd ed. New York; Berlin, Germany; Vienna, Austria: Springer-Verlag, 2000.
[37] H. R. van der Vaart, "A comparative investigation of certain difference equations and related differential equations: Implications for model-building," *Bull. Math. Biol.*, vol. 35, no. 1–2, pp. 195–211, Feb. 1973, doi: 10.1007/BF02558806.
[38] D. T. Mihailović and G. Mimić, "Kolmogorov complexity and chaotic phenomenon in computing the environmental interface temperature," *Mod. Phys. Lett. B*, vol. 26, no. 27, Oct. 2012, Art. no. 1250175, doi: 10.1142/S0217984912501758.
[39] V. Kreinovich and I. A. Kunin, "Kolmogorov complexity and chaotic phenomena," *Int. J. Eng. Sci.*, vol. 41, no. 3–5, pp. 483–493, Mar. 2003.
[40] R. Courant, K. Friedrichs, and H. Lewy, "On the partial difference equations of mathematical physics," (in German), *Math. Ann.*, vol. 100, no. 1, pp. 32–74, Dec. 1928, doi: 10.1007/BF01448839.
[41] "The Nobel Prize in Physics 2021." *NobelPrize.org*. https://www.nobelprize.org/prizes/physics/2021/summary/ (accessed Dec. 27, 2021).
[42] D. Garisto. "Why the Physics Nobel honored climate science and complex systems." *Scientificamerican.com*. https://www.scientificamerican.com/article/why-the-physics-nobel-honored-climate-science-and-complex-systems/ (accessed Dec. 30, 2021).
[43] Ž. Saramago, *Udvojeni čovek*, A. Kuzmanović Jovanović, Trans., Belgrade, Serbia: Laguna, 2015.
[44] "The Nobel Prize in Chemistry 1977." *NobelPrize.org*. https://www.nobelprize.org/prizes/chemistry/1977/summary/ (accessed Dec. 27, 2021).
[45] "Complex time—adaptation, aging, arrow of time." *Santafe.edu*. https://www.santafe.edu/research/themes/complex-time (accessed Dec. 27, 2021).
[46] G. Mengaldo, A. Wyszogrodzki, M. Diamantakis, S. J. Lock, F. X. Giraldo, and N. P. Wedi, "Current and emerging time-integration strategies in global numerical weather and climate prediction," *Arch. Comput. Methods Eng.*, vol. 26, no. 3, pp. 663–684, Jul. 2019, doi: 10.1007/s11831-018-9261-8.
[47] H. Wan, P. J. Rasch, M. A. Taylor, and C. Jablonowski, "Short-term time step convergence in a climate model," *J. Adv. Model. Earth Syst.*, vol. 7, no. 1, pp. 215–225, Feb. 2015, doi: 10.1002/2014MS000368.

[48] A. Tychonoff, "About the topological expansion of spaces," (in German), *Math. Ann.*, vol. 102, no. 1, pp. 544–561, Dec. 1930.

[49] J. Gunawardena, "Theory and mathematical methods," in *Comprehensive Biophysics*, vol. 9, E. H. Egelman, Ed., Amsterdam, The Netherlands; New York, NY, USA: Elsevier, 2012, pp. 243–267.

[50] L. Michaelis and M. Menten, "The kinetics of invertase action," (in German), *Biochem. Z.*, vol. 49, pp. 333–369, 1913.

[51] T. L. Parsons and T. Rogers, "Dimension reduction for stochastic dynamical systems forced onto a manifold by large drift: A constructive approach with examples from theoretical biology," *J. Phys. A Math. Theor.*, vol. 50, no. 41, Sep. 2017, Art. no. 415601, doi: 10.1088/1751-8121/aa86c7.

[52] T. P. Lolaev, "Time as a function of a biological system," in *Space-Time Organization of Ontogenesis*, T. P. Lolaev, Ed. Moscow, Russia: Moscow State University, 1998, pp. 30–35.

[53] D. T. Mihailović and I. Balaž, "Forming the functional time in process of biochemical substance exchange between cells in a multicellular system," *Mod. Phys. Lett. B*, vol. 26, no. 16, Jun. 2012, Art. no. 1250099, doi: 10.1142/S0217984912500996.

4

Are Model and Theory Synonymous in Physics? Between Epistemology and Practice

4.1 Some Background Concepts and Epistemology

It seems that views on the relationship between theory and model in science, art, and everyday communication can be summarized as follows: Theory and model are closely related terms, but there is a fundamental difference between them. The theory is a conceptualized framework for a phenomenon, while a model is a physical representation of a concept which is intended to make this concept more understandable. Thus, the theory provides answers to various problems established in science, and models can be considered representations created to explain the theory. We intend to define the relationship between theory and model in the physics of complex systems. Perhaps the best starting point is Duhem's book *The Aim and Structure of Physical Theory* [1], written by a solid scientist and connoisseur of the philosophy of physics more than a century ago, and his opinion on physical theories: "A physical theory is not an explanation; it is a system of mathematical propositions whose aim is to represent as simply, as completely, and as exactly as possible a whole group of experimental laws" [1]. Duhem adheres resolutely to the separation of physics from metaphysics, which shows our inability to reach the deeper levels of reality. According to him, only physical theories based on the relations of an algebraic nature (algebraic can probably be read as meaning "mathematical") between phenomena can exist. This approach is supported by many physicists (including quantum physicists) but is also considered narrow by some of them because it excludes what is necessary for the further progress of physics: a complex structure of real problems and phenomena. Duhem strongly opposes mechanical models: "Those who are commissioned to teach engineering are therefore eager to adopt the English methods and teach this sort of physics, which sees even in mathematical formulas nothing but models" [1]. Analyzing Duhem's understanding of theory, one gets the impression that theory and model can be used as synonyms. However, he does not mention the word model, which is, in our opinion, incorporated in the statement mentioned above. Duhem's understanding of theory and model allows this set of mathematical formulations (i.e., a model) to be simpler than others for describing a particular phenomenon. For example, the propagation of the temperature wave in a porous material is described by the partial differential equation $C(\partial T/\partial t) = (\partial/\partial z)(\lambda \partial T/\partial z)$, where C is the volume-specific weight of the porous material, and λ is its thermal conductivity. The assumption that the thermal diffusivity $k = \lambda/C$ is constant leads to the equation for the temperature diffusion $\partial T/\partial t = k\partial^2 T/\partial z^2$ that is easier to solve in climate models. Although this assumption is physically unrealistic, it gives "reasonable" results accepted by climate modeling scientists. In other words, "essentially all models are wrong, but some are useful" [2], as George E.P. Box, a British statistician, observes.

While theory (i.e., a model) *describes* phenomena in physics, the concept of the model is based on *interpretation* in mathematics. Figure 4.1 shows that if

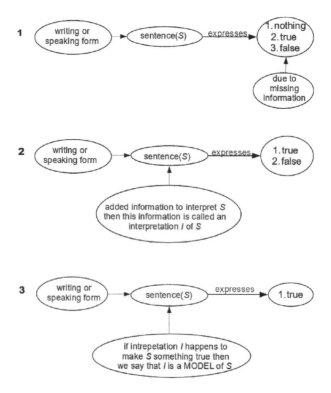

FIGURE 4.1
The flow diagram of the definition of a model in mathematics.

interpretation makes something true, then we say that *I* is a model of *S*. In mathematics, the connection between formal theories and their models is the matter of model theory, which is the part of mathematical logic. In the last half of the twentieth century, the model theory was linked to philosophy through its discipline, the philosophy of mathematical practice.

Mathematics is *conditio sine qua non* for establishing a scientifically acceptable theory. Physicists have recognized similarities between some mathematical relations and relations they have observed while researching physical phenomena. Mathematical physics deals with developing mathematical methods for physical problems (this has also improved some mathematical methods). The contribution of mathematics to physical theories is impressive: (1) *Newtonian gravity* (general relativity), (2) *classical electricity and magnetism* (i.e., classical field theory) (special relativity), (3) *classical mechanics* (special relativity, quantum mechanics), (4) *quantum mechanics* (classical mechanics, quantum field theory), (5) *general relativity* (special relativity, quantum field theory, "string theory"), (6) *quantum field theory* (classical mechanics, classical field theory, quantum field theory, "string theory"), (7) *special relativity* (classical mechanics, classical electricity and magnetism, quantum field theory, general relativity), and (8) *string theory* (quantum field theory, general relativity). String theory, which studies the idea of replacing all the particles of matter and force with just one element—tiny vibrating strings, has been evolving toward its explosion for the last fifty years [3]. Although without impressive progress, this theory has profoundly impacted the relationship between mathematics and physics, even though some physicists disagree. Scientists have broadly pointed out that complex systems modeling in physics needs new mathematical approaches, or "new mathematics." It seems that the part of that mathematics is also "physical mathematics" that has been changing into a separate unit, slowly driven by problems in physics, quantum gravity, string theory, and supersymmetry.

A complex system implicitly contains many properties, although a complete definition of any complex system may not be derived. Then the question that arises is how to model a complex system. One possible answer is that "complex systems modelling is defined by the application of diverse mathematical, statistical and computational techniques, to generate insight into how some of the most complicated physical and natural systems in the world function" [4]. This definition is rather literal than the definition that has a fundamental meaning. We do not have an exact answer to how a model is defined in the physics of complex systems since this answer perhaps does not exist if we consider Gödel's incompleteness theorems and Barrow's comment in subchapter 1.4. However, what is certain is that we face many obstacles to the modeling of physical systems. Namely, the number of *strategies* for a physical system is undoubtedly limited by physical laws and the fact that the concept of adaptation does not exist within that system (the "tyranny" of physical laws, as Stephens notes [5]).

4.2 Choice in Model Building

Considering model choice and building from Duhem's perspective, the focus is to bring choice and building as close as possible to the theory's description. Certain questions in complex systems modeling in physics cannot be answered because they are challenging to answer or, *per definitionem*, have no answer. We discuss some of them, including questions inherent to complex physical systems. Let us begin with general principles. One of the first decisions is whether a model should be *macroscopic* or *microscopic*. If we analyze a phenomenon occurring at a microscopic level, the model needs to be microscopic. If the phenomenon is macroscopic, the model can be either macroscopic or microscopic, depending on the phenomenon. Closely related to this division is the one into *continuous* and *discrete* models. Macroscopic models that use basic equations to capture main microstructural features are always continuous, while microscopic models are discrete. A discrete model is transformed into a continuous one after averaging (i.e., a statistical treatment), which is the basis of classical electrodynamics. Likewise, solid-state physics leads to the physics of crystals. Ultimately, one of physics' objectives is to determine macroscopic properties from microscopic models. The following vital point is that *symmetry* must be included in the model. For example, the symmetry of a crystal produces much information about physical properties, so it must be integrated from the beginning. In the study of phase transitions, an important approach is to construct the free energy of a system (a scalar) from the components of the order parameter by using the symmetries of a material. The problem becomes more complicated when we construct models based on hypotheses about certain particles or field quanta. We usually do not know the types of interactions between involved objects, so we base our idea on symmetries. We start with free space homogeneity and isotropy symmetries and time—homogeneity. Time anisotropy (i.e., the existence of a privileged direction—the arrow of time) is discussed in detail in subchapter 3.1. To relate our model to experimental results, we introduce some abstract symmetry properties and require that the model satisfies those assumptions (quantum field theory is an example). It is common practice for physicists to build the model in two steps. We first consider a system with its intrinsic symmetries and then examine interactions between its constituents or an external agent possessing its own symmetry. The interesting effects that appear in that moment are well described with the expression "C'est la dissymétrie qui crée le phénomène" (meaning "It is the asymmetry that creates a phenomenon") written by Pierre Curie in [6] in 1894.

One important thing to consider is the order of the magnitude of quantities. Let us suppose that we know from experimental data or general principles that some quantities have larger values than others (this comparison is only correct if all quantities are expressed in dimensionless units). In that case, a small parameter takes on the value of the inverse of the largest quantity. Once we include the small parameter in the model, we can simplify calculations in several ways. The most widely used method for this purpose is the theory of perturbations, in which a quantity is represented as a power series in terms of the small parameter, although its convergence is often questionable. Further, involved approximations are sometimes treated as different models. Finally, we often encounter the problem of the analytical expression for the dependence between two quantities. The most suitable empirical guess for fitting data is *the power law*—that is, one quantity (variable) can be represented as a certain power of the other quantity (variable). The power is a number, positive or negative, integer, or fraction; most empirical models are based on it.

Complex systems may be defined as those with many degrees of freedom and nonlinear interactions. Still, these conditions are satisfied by all natural phenomena, so they are not sufficient assumptions for analyzing complex systems. For example, the edge of chaos. It is a space in which degrees of freedom lie between complete order and complete chaos. It exists within various complex systems, but Stephens [5] says that the edge of chaos cannot differentiate between complex and noncomplex forms. Therefore, it is a characteristic of complex systems but not a defining one. Although experience tells us so, not everything is covered by the power law in nature and the physical world. There are various examples of effective degrees of freedom with characteristic scales—objects in the macroscopic world that range from massive elementary particles to macromolecules. However, the fact that the effective degrees of freedom differ significantly between characteristic scales influences the choice of a model. In addition, such effective degrees of freedom show a hierarchical connection.

Do we model complexity or complex system? We can understand this issue more if we consider the illustration of a simple mathematical model [7] for a group of point particles

$$d_i = -\sum_{j \neq i} \frac{c_j(t) - c_i(t)}{|(c_j(t) - c_i(t))|} + \sum_{j=1}^{n} \frac{v_j(t)}{|v_j(t)|} \qquad (4.2.1)$$

where c_i are position vectors, v_i are velocities, and $\hat{d}_i = d_i / |d_i|$ is the direction of the force that occurs because of the interaction of *the* i-th particle with other particles. The terms on the right-hand side are the repulsion and attraction between particles, respectively. The force that aligns the direction of a particle with d_i is generated stochastically by adding a small random number. This simple model for point particles cannot separate the complex from noncomplex systems. There is nothing to suggest, such as the hierarchy of the effective degrees of freedom or a strategy, a more meaningful approach to describing an element of a complex system. This model is applied effectively to modeling the dynamics of fish shoals [7] that can be, without doubt, attributed to complex systems. It is natural to ask ourselves how a simple equation can model a complex system. And why do we model a complex system but not its most discriminating feature—complexity [5]?

The relationship between complex systems and the host environment is crucial for the choice of a model and its building. The internal state of a system can change dynamically, influenced by changes in the surrounding environment [8]. The number of different strategies determines dynamical rules for updating the system's status, an important characteristic of biological complex systems (meaning that they can adapt to a new updated rule), or the strategy within a hostile environment. What is the situation in physics concerning this strategy? Two central problems are as follows: 1. The correspondence between the surrounding environment and the model. 2. Each level of describing complex systems may be autonomous without any connection with other levels. Let us consider molecules and atoms with a low ionization potential in contact with a heated surface. Thermal conditions cause molecules to dissolve; thus, this environment is unsuitable for molecules, so they cannot survive or adapt. Similarly, atoms start dissociating by losing electrons when we raise the temperature. In conclusion, physics is forever getting caught between microscopic and macroscopic states. A deeper understanding of a complex physical phenomenon can provide a new update of internal rules and, as such, be a further strategy in physics.

4.3 The Discrete Versus Continuous Dichotomy: Time and Space in Model Building

Discrete and continuous behavior and the corresponding strategies for building complex systems models in physics are deeply interrelated. Lesne [9] offers comprehensive but concise comments on "the inevitable dance" between them:

> A key point is that *the discrete is not an approximation of the continuum nor the converse*. Great care should be taken when passing from discrete to continuous models and conversely, since their natures are irremediably different. Paradoxes and inconsistencies between discrete and continuous viewpoints only appear when we forget that our descriptions, and even physical laws, are only *idealized abstractions, tangent to reality in an appropriate scale range*, and unavoidably bounded above and below.

Both exist in any physical phenomenon and signify a fundamental dichotomy in mathematics [10]. Let us take into consideration the connotation of continuous in physics and mathematics. It is less equivocal in physics than in mathematics because it relates only to the time and smoothness of *continuous dynamical systems*. In physics, this relation occurs in time and space: 1. *Discrete versus continuous in time*. We first look at the relationships between discrete and continuous time and try to bridge their differences in theoretical modeling, data analysis, and numerical implementation. One of the discretization procedures is the *Poincaré section discretization method* [11]. Let S be an autonomous differentiable dynamical system generating a flow with a periodic orbit O of period T. The successive intersections $x_0, x_1, x_2...$ of a continuous trajectory with the hypersurface X define the return map φ_X (the *Poincaré map*), where $\varphi_X(x_0) = x_1, \varphi_X(x_1) = x_2,...$ The time when the intersection occurs is called the return time [12], and the discretization step is not chosen arbitrarily since dynamics already specifies it. Apart from the *Poincaré map*, there are other procedures: a. The *Birkhoff map* [13] in which the knowledge of the pre-collision state determines the dynamical recursion. b. *Discrete models* that can be introduced directly and are used in population dynamics. c. *Euler's discreditation scheme* that is widely popular in physics. The theory of dynamical systems in physics represents a mathematical basis for the time evolution of a phenomenon. Systems are specified in the form of physical laws that generate evolution equations. Its parameters describe the current state of the environment and can change over time in an autonomous dynamical system. A difference equation $x_{n+1} = f(x_n), n \in Z^+$, where $f: R^d \to R^d$ is a first-order autonomous difference equation on the state space R^d(Z^+ denotes nonnegative integers). Restriction to linear equations is not a loss of generality, so there is no loss of generality in the restriction to the first-order equations since higher-order difference equations can be reformulated by using an appropriate higher dimensional state space.

The successive iteration of an autonomous difference equation generates the forward solution mapping $\pi: Z^+ \times R^d \to R^d, x_n = \pi(n, x_0) = f^n(x_0) = \underbrace{f \circ f \circ \cdots \circ f(x_0)}_{n\times}$ that satisfies the initial condition $\pi(x_0) = x_0$ and the semigroup property $x_n = \pi(n, \pi(m, x_0)) = f^n(\pi(m, x_0)) = f^n \circ f^m = f^{n+m}(x_0) = \pi(m+n, x_0)$ for all $m, n \in Z^+, x_0 \in R^d$. Equation $x_{n+1} = f(x_n)$ can be solved numerically by using the known initial condition to step either forward $x_{n+1} = x_n + \Delta \tau f(x_n)$ or backward $x_{n+1} = x_n + \left(\Delta \tau / \left(1 - \Delta \tau \frac{\partial f}{\partial x}(x_n) \right) \right) f(x_n)$ in time, where $\Delta \tau$ is the time step, and n is the time iteration. 2. *Discrete versus continuous in real space*. The three-dimensional ($d = 3$ space (or higher dimension) may be seen either as a continuum in the Euclidian space filled with geometric objects with the same dimensions that determine the position or as a lattice represented with a series of points arranged in a different pattern, where d integers characterize the position. Lattice models are effective and reliable for analyzing the universal properties of physical systems having the same symmetries and geometrical properties. To see an object in the macro or micro world, one must have appropriate "optics" through which the observed object seems to be intrinsically discrete or even isolated (individual atoms can be seen under the UK's super STEM electron microscope). For example, the atom is delocalized at small scales (those of quantum mechanics). Because a probability distribution determines the position of a particle, we cannot say that discrete objects offer a more objective perspective than an arbitral division of space into cells [9]. 3. *Discrete versus continuous in a phase space*. The system state can be represented by a continuum (vector space) or varies within a finite or countable set of configurations. The relationship between agent-based and kinetic continuous state descriptions has taken place in complex systems modeling: population dynamics, granular media, chemical kinetics at different scales, etc. 4. *Discrete versus continuous in conjugate space* is considered in the context of spectral analyses since spectra offer another modality to a dichotomy. In spectral analyses, the underlying real space is split into cells (conjugate space), so they are powerful methods for decomposing physical behavior into elementary components. This method is mainly applied when we want a richer picture of the system's behavior. Spectra include frequencies (or time periods), wave vectors (or wavelengths), energy levels, correlation times, and amplification rates.

We close this subchapter with some comments on digital computing. An accuracy limitation bounds this process. There is a contrary variation between the *method error* (i.e., the discretization error) and the *round-off error* against the step size whose chosen value regulates 1. the stability of a model and its efficiency; and 2. computational time. Thus, the *computational uncertainty principle* is similar to the well-known Heisenberg uncertainty relation. The mathematical expressions of the principle are $\Delta e + \Delta \tilde{r} \geq C$ and $\Delta e \cdot \Delta \tilde{r} \geq \sigma$. Δe is a measure of uncertainty attributed to the limitation of a numerical method itself, $\tilde{r}$ is a measure of uncertainty resulting from the limitation of computer accuracy, C and σ are positive numbers dependent on differential equations, and machine precision is finite [14]. If we relate this principle to discretization and the round-off error, the adjoint variable is greater if one error is smaller than the other. The choice of step size is crucial for numerical computations, especially in their uses for solving partial differential equations. Indeed, the numerical method places upper limits on the size of the step, while the number of integration steps is limited by finite computer accuracy.

4.4 The Predictability of Complex Systems. Lyapunov and Kolmogorov Time

The predictability of a complex system usually refers to 1. the time evolution of the system from which we can obtain information; and 2. the content of obtained information. When predictability is considered, our attention is mainly on a macroscopic model that predicts the system's state for a longer time and spatial scale. We speak about predictability in the context of a microscopic model if we evaluate how much the final state depends on the information in the initial state and the system's history.

A complex system in physics includes numerous elements that are important but only sometimes sufficient for its functioning (see subchapter 4.3). The questions are what it means to model a complex physical system and whether obtaining information from that model is possible. It appears that these questions have no straightforward answers. Some features of components in complex systems cannot be measured successfully, which complicates model building, and even simple models produce different results. Physicists often use shorter computational methods to obtain results. This practice is only possible when 1. computations are more sophisticated than those performed by physical systems; 2. a physical system behaves like a computer, known as the *computational irreducibility* introduced in [15]. It should be noted that a direct simulation or observation can only explain the system's behavior; predictability based on a general procedure is not possible. For example, computationally irreducible physical processes and even computationally reducible ones at a coarse-grained description level can be predictable [16]. The predictability of disordered spin systems that follow a deep quench is a complex issue. Long-term predictability in mean-field models fully determines the final state. In contrast, predictability can seriously be reduced in short-range models [17].

The greatest and most visible breakthrough in physical complex systems modeling was made in the interaction between the earth and the atmosphere in meteorology. These models, at different spatial and temporal scales, have also opened up new horizons in other sciences; therefore, scientists take a huge interest in their improvements to obtain better predictability. Lorenz [18] analyzes the outcomes of the integration of a set of 12 ordinary differential equations from a simplified two-layer baroclinic model of the atmosphere. He concludes that the solutions of the model are not necessarily stationary nor periodic (this was a prevailing opinion at the time). Furthermore, he obtains solutions that vary without evidence of periodicity. Lorenz establishes the theoretical basis of weather and climate predictability (in Lorenz's sense) in the theory of error growth [19]. For an N-dimensional system, if $x(t)$ is a small error, then the error growth from t_0 to t_1 is given by $x(t) = E(t_0, t_1)x(t_0)$, where E is a square matrix depending on nonlinear equations in a model between t_0 and t_1. Supplementarily, it was shown that the values of errors lie inside a small sphere with the radius ϵ that evolves into an ellipsoid whose semiaxes are $\epsilon\delta_i$, where $\delta_1, ..., \delta_N$ are the singular values of E or the square root of the eigenvalues of EE^T, with E^T as the transpose of E. If singular values are arranged in decreasing order, the error growth occurs if $\delta_1 > 1$. The eigenvectors of EE^T represent the orientation of errors. During a certain time interval, errors grow or decay depending on the magnitude of δ_i. The overall growth of small errors can be determined by considering the interval $(t - t_0)$

to be large. The limiting values defined as $\lambda_i = \left(1/(t-t_0)\right)\lim_{t\to\infty}\ln\delta_i$ are called the Lyapunov exponents. λ_i are generally independent of the initial state for many systems and represent average exponential growth or decay. A similar definition for the eigenvectors of EE^T in the limit of the large time interval yields the Lyapunov vectors used in the predictability analysis and as initial perturbations for ensemble forecasting in numerical weather prediction. In short, the predictability of current weather forecasting and climate models is limited by two issues: 1. Reliable forecasts cannot be made for more than ten days. 2. Instantaneous states in climate models cannot be predicted, and predictions are only possible for some aspects of climate variability [20]. In addition, chaos in deterministic nonlinear systems can occur in vertical and horizontal fluxes exchange at the environmental interfaces of the earth-atmosphere system [8].

We also point out a time scale called the *prediction horizon*—the Lyapunov time $t_{\text{lyap}} = 1/\lambda_{\max}$ (expressed in the units of recorded series), where $\lambda_{\max}$ is the largest positive Lyapunov exponent. It is a period after which a dynamical system becomes unpredictable and enters a chaotic state, indicating the limits of predictability. If t_{lyap} increases when $\lambda \to 0$, then accurate long-term predictions are possible (the upper panel of Figure 4.2). Research suggested that t_{lyap} overestimates the actual value of the period. To correct this overestimation, Mihailović et al. [21] introduce the *Kolmogorov time* $t_{kol} = 1/KC$, where KC is Kolmogorov complexity. This time quantifies the time window size within which complexity remains unchanged. Hence, the presence of a narrow window t_{kol} significantly reduces the length of the effective prediction horizon (the lower panel of Figure 4.2).

4.5 Chaos in Environmental Interfaces in Climate Models

From the beginning of the research on climate models, there have been many debates about what models should include, how reliable they are, and issues related to chaos. Opinions on the complexity of climate models are different and narrow because the extravagance and way they are presented distract the scientific community from more fundamental issues. In this subchapter, we consider climate models in the context of their modeling and as highly complex systems that consist of several major components: the atmosphere, the hydrosphere, the cryosphere, the land surface with the biosphere, and the interfacial fluxes between these components. According to observations, the climate system exhibits chaotic behavior at all time scales [22].

Are Model and Theory Synonymous in Physics?

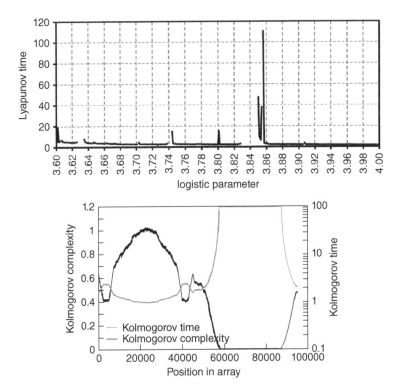

FIGURE 4.2
Lyapunov time (*upper*) and Kolmogorov time (*lower*) for the logistic equation. Empty intervals indicate $t_{lyap} < 0.$; the logistic parameter was $r = 3.75 + 0.25 \sin((2\pi)i/N)$, where i is the position in time, N is the size of the time series; the running Kolmogorov complexity was calculated using the window of 5000 samples long. (Reproduced by permission from [21].)

Finite systems of deterministic ordinary nonlinear differential equations may be designed to represent forced dissipative hydrodynamics flow. Solutions of these equations can be identified with trajectories in phase space. For those systems with bounded solutions, *it is found that non-periodic solutions are ordinarily unstable with respect to small modifications, so that slightly differing initial states can evolve into considerably different states. Systems with bounded solutions are shown to possess bounded numerical solutions.* A simple system representing cellular convection is solved numerically. *All of the solutions found to be unstable, and almost all of them are nonperiodic.* The feasibility of very long-range weather prediction is examined in the light of these results [23].

Lorenz's paper "Deterministic Nonperiodic Flow" [23], which formed the basis for chaos theory, was one of the most outstanding achievements in physics in the twentieth century. However, few scientists noticed that at the time. Lorenz delivered the lecture "Predictability: Does the Flap of a Butterfly's Wings in Brazil Set off a Tornado in Texas?" [24] in 1972, but it was unnoticed until Gleick's book *Chaos: Making a New Science* [25] made the term "butterfly effect" and Lorenz's discovery popular fifteen years later.

It is accepted that weather is in a state of deterministic chaos caused by nonlinearities in the Navier–Stokes equations. A system may only be predictable for up to seven days because of its sensitivity to initial conditions. Chaos in climate models occurs for the same reasons; however, a significant difference is that its source is the linkage of different subsystems that produces something more complex compared with deterministic chaos in weather models. Weather and climate models are identical in principle. Climate models function because weather is chaotic and not random. When the climate model is integrated over a longer time period, climate emerges from chaos. Here we discuss the chaos in the energy exchange at *environmental interfaces* [8] between land and the atmosphere in climate models.

In climate models, the soil surface–air layer can be treated as an environmental interface (i.e., a dynamical system), also sensitive to initial conditions. In such a system, chaotic fluctuations occur while we calculate the environmental interface temperatures in the soil surface and deeper soil layer via two partial differential equations [26]. Depending on the climate model grid cell and atmospheric conditions at a reference level, these equations can be written as [27, 28]

$$x_{n+1} = rx_n(1-x_n) + \varepsilon y_n \quad (4.2.2a)$$

$$y_{n+1} = \varepsilon(x_n + y_n) \quad (4.2.2b)$$

FIGURE 4.3
Cover type in the climate model grid cell that can determine the occurrence of chaos. *Spinifex grassland*, Yakabindi station, Western Australia (a). (Image courtesy of Vilis Nams.) (b) Patterns of *P. bulbosa* observed in the Northern Negev (b). (Reproduced by permission from [29].)

Are Model and Theory Synonymous in Physics?

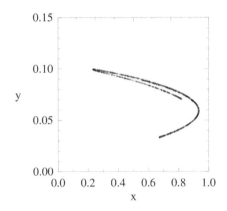

FIGURE 4.4
The attractor in the phase space of the coupled system (4.2.2) for $r = 3.7$ and $\varepsilon = 0.1$ with the initial conditions $x_0 = 0.2$ and $y_0 = 0.4$. (Reproduced by permission from [28].)

where x_n and y_n are the dimensionless environment interface temperatures at the soil surface and deeper soil layer, respectively; r is the coefficient, and ε is the coupling parameter, where $r \in [0,4]$ and $\varepsilon \in [0,1]$. This system is observed when 1. the model grid cell consists of patches similar to the one shown in Figure 4.3; 2. the time step is not chosen properly because of limitations by the computational uncertainty (see subchapter 4.3).

The occurrence of an attractor in the system (4.2.2) is possible for the small values of the coupling parameter ε in a phase space (Figure 4.4). In this case, the system (4.2.2) is close to Hénon's map; thus, the attractor is similar toHénon's.

References

[1] P. Duhem, *La théorie physique son objet et sa structure*, 2nd ed. Paris, France: Chevalier et Rivière, 1914.
[2] R. Wasserstein, "George Box: A model statistician," *Roy. Statist. Soc.*, vol. 7, no. 3, pp. 134–135, Sep. 2010, doi: 10.1111/j.1740-9713.2010.00442.x.
[3] G. Veneziano, "Construction of a crossing-symmetric, Regge-behaved amplitude for linearly rising trajectories," *Nuovo Cimento A*, vol. 57, no. 1, pp. 190–197, Sep. 1968, doi:10.1007/BF02824451.
[4] "Complex systems modelling." *Sheffield.ac.uk*. https://www.sheffield.ac.uk/dcs/research/groups/complex-systems-modelling (accessed Feb. 24, 2022).
[5] C. R. Stephens, "What isn't complexity?," 2015, *arXiv*:1502.03199.

[6] P. Curie, "On symmetry in physical phenomena, symmetry of an electric field and a magnetic field," (in French), *J. Phys. Theor. Appl.*, vol. 3, no. 1, pp. 393–415, 1894, doi:10.1051/jphystap:018940030039300.

[7] I. D. Couzin, J. Krause, N. R. Franks, and S. A. Levin, "Effective leadership and decision-making in animal groups on the move," *Nature*, vol. 433, no. 7025, pp. 513–516, Feb. 2005, doi:10.1038/nature03236.

[8] D. T. Mihailović, I. Balaž, and D. Kapor, *Time and Methods in Environmental Interfaces Modelling: Personal Insights* (Developments in Environmental Modelling 29). Amsterdam, The Netherlands; New York, NY, USA: Elsevier, 2016.

[9] A. Lesne, "The discrete versus continuous controversy in physics," *Math. Struct. Comput. Sci.*, vol. 17, no. 2, pp. 185–223, Apr. 2007, doi:10.1017/S0960129507005944.

[10] J. Franklin, "Discrete and continuous: A fundamental dichotomy in mathematics," *J. Humanist. Math.*, vol. 7, no. 2, pp. 355–378, Jul. 2017, doi:10.5642/jhummath.201702.18.

[11] M. H. Poincaré, *Les méthodes nouvelles de la mécanique céleste*. Paris, France: Grauthier-Viltars, 1892.

[12] J. Guckenheimer and P. Holmes, *Nonlinear Oscillations, Dynamical Systems, and Bifurcations of Vector Fields (Applied Mathematical Sciences 42)*. New York; Berlin, Germany; Vienna, Austria: Springer-Verlag, 1983.

[13] P. Gaspard, "Maps," in *Encyclopedia of Nonlinear Science*, A. Scott, Ed., Evanston, IL, USA: Routledge 2005, pp. 548–553.

[14] L. Jianping, "Computational uncertainty principle: Meaning and implication," *Bull. Chin. Acad. Sci.*, vol. 15, no. 6, 2000.

[15] S. Wolfram, "Undecidability and intractability in theoretical physics," *Phys. Rev. Lett.*, vol. 54, no. 8, pp. 735–738, Feb. 1985, doi:10.1103/PhysRevLett.54.735.

[16] N. Israeli and N. Goldenfeld, "Computational irreducibility and the predictability of complex physical systems," *Phys. Rev. Lett.*, vol. 92, no. 7, Feb. 2004, Art. no. 074105, doi:10.1103/PhysRevLett.92.074105.

[17] J. Ye, R. Gheissari, J. Machta, C. M. Newman, and D. L. Stein, "Long-time predictability in disordered spin systems following a deep quench," *Phys. Rev. E*, vol. 95, no. 4, Apr. 2017, Art. no. 042101, doi:10.1103/PhysRevE.95.042101.

[18] E. N. Lorenz, "The statistical prediction of solutions of the dynamic equations," in *Proc. Int. Symp. Numer. Weather Predict.*, Tokyo, Japan, Nov. 1960, pp. 629–635.

[19] E. N. Lorenz, "A study of the predictability of a 28-variable atmospheric model," *Tellus*, vol. 17, no. 3, pp. 321–333, 1965, doi:10.3402/tellusa.v17i3.9076.

[20] V. Krishnamurthy, "Predictability of weather and climate," *Earth Space Sci.*, vol. 6, no. 7, pp. 1043–1056, Jul. 2019, doi:10.1029/2019EA000586.

[21] D. T. Mihailović et al., "Analysis of solar irradiation time series complexity and predictability by combining Kolmogorov measures and Hamming distance for La Reunion (France)," *Entropy*, vol. 20, no. 8, p. 570, Aug. 2018, doi:10.3390/e20080570.

[22] J. A. Rial et al., "Nonlinearities, feedbacks and critical thresholds within the earth's climate system," *Climatic Change*, vol. 65, no. 1–2, pp. 11–38, Jul. 2004, doi:10.1023/B:CLIM.0000037493.89489.3f.

[23] E. N. Lorenz, "Deterministic nonperiodic flow," *J. Atmos. Sci.*, vol. 20, no. 2, pp. 130–141, Mar. 1963, doi:10.1175/1520-0469(1963)020<0130:DNF>2.0.CO;2.

[24] E. N. Lorenz, "Predictability: Does the flap of a butterfly's wings in Brazil set off a tornado in Texas?" (lecture, American Association for the Advancement of Science, Dec. 29, 1972).
[25] J. Gleick, *Chaos: Making a New Science*. New York, NY, USA: Viking, 1987.
[26] D. T. Mihailović and G. Mimić, "Kolmogorov complexity and chaotic phenomenon in computing the environmental interface temperature," *Mod. Phys. Lett. B*, vol. 26, no. 27, Oct. 2012, Art. no. 1250175, doi:10.1142/S0217984912501758.
[27] G. Mimić, D. T. Mihailović, and M. Budinčević, "Chaos in computing the environmental interface temperature: Nonlinear dynamic and complexity analysis of solutions," *Mod. Phys. Lett. B*, vol. 27, no. 26, Oct. 2013, Art. no. 1350190, doi:10.1142/S021798491350190X.
[28] G. Mimić, "Nonlinear dynamical analysis of the physical processes in the environment," Ph.D. dissertation, Dept. Physics, Univ. Novi Sad, Novi Sad, Serbia, 2016.
[29] I. Bordeu, M. G. Clerc, P. Couteron, R. Lefever, and M. Tlidi, "Self-replication of localized vegetation patches in scarce environments," *Sci. Rep.*, vol. 6, Sep. 2016, Art. no. 33703, doi:10.1038/srep33703.

5
How Can We Assimilate Hitherto Inaccessible Information?

5.1 The Physicality, Abstractness, and Concept of Information

Communication is the transfer of information from one point (a place, person, or group) to another. It involves at least one sender, a message, and a recipient. In terms of a medium, it can be divided into several distinctive categories: (1) spoken or verbal communication; (2) nonverbal communication; (3) written communication; and (4) visualization. If we look at these categories, we can ask ourselves a question: Is information physical or abstract [1]? *Physicality* implies that information has a physical manifestation or physical existence. Landauer [2] comes to the following conclusion:

> Information is inevitably inscribed in a physical medium. It is not an abstract entity. It can be denoted by a hole in a punched card (notwithstanding this long-abandoned technology it is still noticeable as an example), by the orientation of a nuclear spin, or by the pulses transmitted by a neuron. The quaint notion that information has an existence independent of its physical manifestation is still seriously advocated. This concept, very likely, has its roots in the fact that we were aware of mental information long before we realized that it, too, utilized real physical degrees of freedom.

On the contrary, abstractness assumes that information is an idea that exists in our minds. Timpson [3] argues that information is an abstract entity:

> Information, what is produced by a source, or what is transmitted, is not a concrete thing or stuff. It is not so, because, as we have seen, what is produced/transmitted is a sequence type and types are abstract. They are not themselves part of the contents of the material world, nor do they have a spatiotemporal location.

The important conclusion from this elaboration is that information can be either physical or abstract, but it certainly exists outside its physical manifestation and our ability to detect it. Therefore, almost anything can be perceived as information:

> The phenomenon being observed is information, the pattern of organization of matter and energy as it exists in the universe and in living beings. The fact that we are observing, however, and claiming the objective existence of patterns of organization such as neurally stored memories, does not imply that our understanding or construction of that objective existence is true, complete, correct, or the only possible understanding. Nor does this claim imply that we deny the subjective variations and uniqueness in each individual's perception, extraction, and use of information in their minds and surroundings.
>
> [4]

Various theoretical approaches to the *concept of information* have been developed and argued in detail. They are rooted in different disciplines, such as physics, mathematics, logic, natural sciences, and philosophy, and gather around two central properties: 1. *Information is extensive*. Extensiveness emerges when we observe and draw conclusions from the surrounding environment. The fundamental concept is *additivity*: the combination of two successive and independent events [5]. For example, let 0.8 be the probability that one will go for a walk and 0.5 the probability that one will receive a call on a cell phone (this happens independently of whether one is at home or not). Then the probability of the realization of both events is 0.40. What about the information included in these independent events? If one is slightly surprised by the event, and another event happens, then the total surprise depends only on the probability of another event. In conclusion, the total information should be the sum of the two independent amounts of information because the corresponding events are also independent. 2. *Information reduces uncertainty*. English empiricists probably first characterized the relationship between uncertainty and information. They observed explicitly that rare events (i.e., events with low probabilities) were the most surprising to us and contained more information. This is formulated mathematically by Hartley [6], who defines the *amount of information* $I(A)$ we obtain when an element is selected from a finite set [7]. The only mathematical function that connects extensiveness and probability and defines information in terms of the negative log of the probability $P(A)$ [7] is $I(A)$:$-\log P(A)$ [8].

In the twentieth century, various proposals to formalize the concept of information were made through qualitative and quantitative information theories. The quantitative ones include some of the following measures: Nyquist's function, Fisher information, the Hartley function, Shannon information, Kolmogorov complexity, entropy measures in physics, Gibbs entropy, etc., and various generalizations, such as Tsallis entropy and Rényi entropy. Note that these entropies are not strictly measures of information. Entropies in physics are associated with the corresponding concepts of information and quantum information, where a qubit is the generalization of the classical bit. A concise list of essential references linked to these theories is available in [7].

The connection between information in an event and the probability of its occurrence entered the physical world in 1872. Ludwig Boltzmann (1844–1906), an Austrian physicist, invented the formula for the entropy S of the system $S = k \log W$, where W is the number of possible microstates consistent with the observable macroscopic states of the system, and k is Boltzmann's constant. From the perspective of information theory, we can interpret this formula as a measure of the amount of entropy in the system, lack of information, or probability of any typical microstate consistent with macroscopic observations. Let us make a comment on the sign in the formula for the amount of information. While W is a number larger than one, the probability of the microstate is obtained by dividing W by the total number of microstates. Since the probability is always equal or less to one, its logarithm is negative, and the sign minus is added in front of the logarithm. We think that the relation $I(A) = -\log P(A)$, which connects two intuitive concepts (i.e., information and probability), is fundamental, such as Duc de Broglie's wave-particle duality, Heisenberg uncertainty relations, or the first law of thermodynamics.

5.2 The Metaphysics of Chance (Probability)

Blaise Pascal (1623–1662) and Pierre de Fermat (1607–1665) are accredited as the fathers of probability since they did the fundamental work on the theory of probability by considering the gambling problem posed by the French writer Antoine Gombaud, Chevalier de Méré (1607–1684) in 1654. Probability is a concept that has contributed significantly to moving the boundaries of science (especially when it is tied to the concept of information). However, it is still one of the "most puzzling" terms for which it is impossible to find the tiniest meaning. It may be understood as a more general question known as the metaphysics of chance: What makes any probability statement true or false? Various interpretations of probability have been aimed at answering this question [9]. We list some of them: classical probability, logical/evidential probability, subjective probability, frequency interpretations, propensity interpretations, and best-system interpretations.

Although often used synonymously, *chance* and *probability* are subtly different terms. The metaphysical aspect of chance is inaccessible to us—that is, our experience and intuition are not sufficient for its understanding, and we want to reach that level of awareness. However, it is still unclear how to approach that "something." Otherwise stated, if we cannot move toward the metaphysics of chance, we cannot move away from it, which is the exact way we understand probability.

Scientists use *probability* (which is the ratio that defines how likely an event is to happen) or *chance* (which does not have any obviousness) to grade events from rare (low probability) to frequent (high probability) [1]. In medical terminology, people increase the probability of contracting cancer by smoking. Still, some events are rare, such as the randomly-rolled cubical dice landing on one of its 12 edges. These events have extremely low probabilities of occurrence and are considered nearly impossible. Imagine that we can throw a dice only two times and obtain six from the first toss and three from the second toss. We might think this result is compatible with the dice being 1. fair—each side can occur with equal probabilities; 2. biased—some sides have a greater chance of coming up. In this example, the metaphysics of chance involves the question: What makes the statement "the probability that the dice will land on its face is 1/6" true? We consider two of the interpretations mentioned above of probability in this subchapter: the classical account of probability and the frequentist account.

According to the classical account of probability [10], probabilities or chances areabstract symmetries that could be determined *a priori* with an equal probability of occurrence. If an experiment results in $n$equally possible outcomes, then we should assign the probability $1/n$ to each of them. Therefore, when a coin is flipped, it can land on either its head or its tail. Each of these outcomes is equally possible, and the probability of both events is 1/2. The problem with Laplace's definition can be expressed through the question: Is it possible that the coin may land on its edge? In this case, Keynes [11] claims that the probability is not 1/2 but 1/3 (the head, face, and edge). He suggests that when one outcome cannot be favored over another, they have an equal probability of occurrence. He calls that principle the *principle of indifference*: "The Principle of Indifference asserts that if there is no *known* reason for predicating of our subject one rather than another of several alternatives, then relatively to such knowledge the assertions of each of these alternatives have an equal probability" [11]. This principle can be contradictory—we almost always obtain the head or the tail when tossing the coin. However, the fact that the coin may land on its edge is completely excluded from our consciousness. Thus, the chance that this relatively simple event will happen remains unnoticed.

Frequentism, another approach, defines probability as the limit of its relative frequency in many trials so that facts about probabilities are *a posteriori*. If we want to determine the probability that the coin will land on its head, we divide the number of times it landed on its head by the total number of outcomes. However, some problems are related to frequentism. First, many events are unrepeatable, and we can obtain only one result. This eliminates other outcomes that are not the result of an experiment. Instead of counting all possible outcomes, only the ones that result from the experiment are counted (this is known as the single case problem). Second, it needs to be defined why some reference classes are more suitable than others for determining the probability of the event. Here we provide an interesting probability analysis of rolling a square cuboidal dice having two squares (the edge y and four rectangle sides [the edges x and y]) (Figure 5.1), focusing on the probability p that a square face is up after rolling the dice.

Mungan and Lipscombe [12] propose a model for calculating this probability by combining probabilistic arguments to produce a formula with a parameter that is fitted from the results of the frequency measurements of rolling the dice with different ratios x/y. The probabilistic arguments include the angle with respect to the ground that defines whether the dice will fall on the rectangle or square side and the ratio of the surfaces on which the dice may fall. Both quantities depend only on the ratio x/y. The probability has the form

$$p = \frac{2}{2+(x/y)^n}\left[1-\frac{2}{\pi}\arctan(x/y)^n\right] \qquad (5.2.1)$$

where n is the parameter determined by fitting experimental data, and it takes a value between 3 and 3.5. In the particular case of $x = y$, the probability is 1/3. In the limiting case of the small x, this can even approximate a coin.

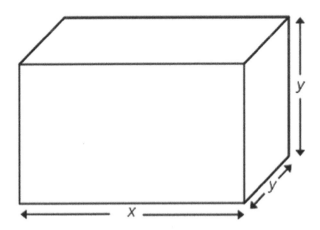

FIGURE 5.1
A cuboidal dice where two opposite faces are squares of edges y, and the other four are rectangles of edges x and y. (Figure courtesy of Miloš Kapor.)

5.3 Shannon Information. The Triangle of the Relationships Between Energy, Matter, and Information

This subchapter describes Shannon entropy, a foundational concept in information theory, along with the relationship between energy, matter, and information. Claude Shannon (1916–2001) became a scientific icon after establishing information theory in his famous paper "A Mathematical Theory of Communication" [8]. His contribution to science was described by one of the outstanding figures in information theory at the turn of the twentieth century, Albert Tarantola [13]: "Shannon must rank near the top of the list of the major figures of 20th century science." The first idea of quantifying information was proposed by Hartley [6]; however, it became clear that information could be defined and measurable after Shannon's paper had been published. Shannon put forward the theory of converting a message generated by a source into a signal and then communicating it to a recipient over a noisy channel. Transmission is corrupted by noise, and the exact reconstruction of the message is difficult to achieve—that is, transmission cannot be done without any errors. Shannon set the limit below which it is possible to transfer the message with zero errors. Above this limit, communication cannot be performed without losing information (known as the noisy-channel coding theorem).

Let X be a discrete random variable on a finite set $X = \{x_1, ..., x_n\}$ with a probability distribution function $p(x) = Pr(X = x)$. Shannon entropy $H(X)$ of X [8] is defined as

$$H(X) = -\sum_{x \in X} p(x) \log_b x. \tag{5.3.1}$$

The logarithm is usually taken to be of the base 2, wherein entropy is measured in bits (one *bit* is the smallest unit of information having a binary value). This entropy is the average information content in a message produced by a source. The general state of a bit is defined by a binary value, while a *superposition* of both defines the general state of a *qubit* [14]. The measurement of the classical bit would not disturb its state, while the measurement of the qubit permanently disturbs the superposition state. By using Shannon's information theory, Vopson [15] estimates the amount of encoded information in all the visible matter in the universe and in each particle at 6×10^{80} bits of information and 1509 bits of information, respectively. Shannon showed that a higher entropy correlates with more information content and vice versa, or the more uncertainty is involved in the transmission, the more information it holds. This is why Shannon entropy is considered a measure of uncertainty apart from being a measure of information.

Let us also mention Shannon's formula in relation to entropy in physics. Since his measure and the thermodynamic entropy in statistical mechanics were mathematically identical, Shannon called it "entropy" [16]. However, those quantities are not the same since Shannon entropy can be defined for any probability distribution, while Boltzmann's entropy is related only to specific probabilities in thermodynamics. We present an interesting story about how Shannon decided to call his measure "entropy."

> My greatest concern was what to call it. I thought of calling it 'information,' but the word was overly used, so I decided to call it 'uncertainty.' When I discussed it with John von Neumann, he had a better idea. Von Neumann told me, 'You should call it entropy, for two reasons. In the first place your uncertainty function has been used in statistical mechanics under that name, so it already has a name. In the second place, and more important, no one knows what entropy really is, so in a debate you will always have the advantage.'
>
> [17]

In physics, entropy originates from the second law of thermodynamics and deals with a set of events. Shannon entropy is a measure of uncertainty in information theory; this self-entropy is associated with a single event. Therefore, information and entropy are fundamentally related since entropy measures the average amount of self-entropies contributing to a system. It can be said that entropy is a measure of the spread of a probability that is often referred to as disorder. Accordingly, the second law of thermodynamics can be understood almost as the absence of the possibility to define precise contexts at a macroscopic level.

Claude Shannon says, "A basic idea in information theory is that information can be treated very much like a physical quantity, such as mass or energy" [18]. From our point of view, he set this quotation in an ontological context—that is, he suggested that information was a real objective trait of reality, such as matter and energy, and not merely a useful mathematical fiction that could assume different forms. In other words, the question is: Can information be reduced to matter and energy and then returned to us as a single element?

Umpleby [19] suggests that these three concepts can physically be connected via the triangle of the relationships between matter, energy, and information. Symbolically, in an MEI triangle, the notations are set as follows: (1) matter (the upper vertex M); (2) energy (the left bottom vertex E); and (3) information (the right bottom vertex I). Note that we replaced his term difference with the term information at the right bottom vertex. In Umpleby's paper, the term "difference" refers to information that is, unlike matter and energy, a function of the observer [19], so it is not an elementary concept. In this book, we use the term "information" following Szilard's consideration of the relationship between information and energy [20].

The triangle sides are represented through the EM side, which is Einstein's relationship $E = mc^2$ and the EI side, due to Szilard, who rigorously demonstrated how physical entropy and information are related in the sense of the theory of communication. That was a breakthrough in the integration of physical and cognitive concepts. Planck's relationship $E = h\nu$ is used for this side when one photon is considered equivalent to one bit [21]. Szilard also recognized that Maxwell's famous demon required information in a sorting process on high- and low-energy particles. He confirmed that the measurement of the velocity of gas molecules would produce more entropy than the sorting process would remove. Thus, this demon was successfully exorcised.
3. MI side represents the so-called *Bremermann's limit* [22], i.e., the physical laws limit the computational rate of any data processing system ("computer"). Its formula was $mc^2/\nu \sim (m/\text{gram})10^{47}$ bits per second, which was later replaced with $c^5/(Gh) \sim 10^{43}$ bits per second, where G is the gravitational constant [23]. From this triangle, we can see two outcomes: 1. "Combining" EM (the relationship between energy and matter) and EI (the relationship between energy and information), we obtain the relationship between information and matter (IM), at least at the atomic level. 2. Considering all the possible information carriers, the relationship between matter and signal is not continuous, strongly depending on the material in which a pattern appears. Thus, up to the atomic level (at which Bremermann's limit applies), we can observe a pattern or set of differences "in molecules (DNA), cells (neurons), organs (the brain), groups (norms), and society (culture)" [24].

Vopson [25] formulates a new principle of the *mass-energy-information equivalence* (MEI triangle), showing that a bit of information has a finite and quantifiable mass while it stores information. Using the mass-energy equivalence principle and $m_{bit}c^2 = k_B T \ln(2)$ (the side connecting M and I triangle vertices), he finds that the mass of a bit of information is $m_{bit} = k_B T \ln(2)/c^2$, where k_B is Boltzmann's constant, and T is the temperature at which the bit of information is stored. At room temperature ($T = 300$ K), the mass of a bit of information has the value of 3.19×10^{-38} kg. Note that the mass-energy-information equivalence principle proposed in this paper strictly applies only to classical digital memory states at equilibrium. Information carried by relativistic media requires another treatment.

5.4 Rare Events in Complex Systems: What Information Can Be Derived From Them?

One famous quote is that information is power in social life and science, which is a cliché. The crucial question is how we can reach *the right* information, particularly in physics. In this attempt, we are aware that 1. Accepting information is a process including physical and cognitive concepts; 2. There are limitations because of Heisenberg's uncertainty principle in physics. Information comes from an event. In modern physics, an *event* is a physical phenomenon that occurs at a specified point in space and time (see subchapter 2.4). However, an event in particle physics refers to the outcomes 1. obtained just after the interaction between subatomic particles; 2. occurring in a very short time duration; 3. that are observed in a well-localized region of space. An event that occurs infrequently is called a *rare event*. According to a probability model, rare events have low probabilities of occurrence [26]. Rare events can usually be of significant interest. The discovery of these events often leads to Copernican turns in physics and, accordingly, in the physics of complex systems. How to obtain information from such events? The computation of the probabilities of rare events is already challenging—analytical solutions are not available for nontrivial problems, while the standard Monte Carlo simulation is computationally ineffective. Therefore, more research efforts focus on using set theory and developing more advanced and efficient stochastic simulation methods. We divide rare events into two categories: *no event registered* and *extremely rare* events.

Extremely rare events ("black swans"). The term "black swan" represents an unpredictable event that causes tectonic changes in nature and society at all scales, including at the global scale (see subchapter 1.1). These events often have catastrophic consequences. In his book *The Black Swan* [27], Naib Nicolas Taleb, the promoter of this term, quotes,

> First, it is an outlier, as it lies outside the realm of regular expectations, because nothing in the past can convincingly point to its possibility. Second, it carries an extreme 'impact.' Third, in spite of its outlier status, human nature makes us concoct explanations for its occurrence after the fact, making it explainable and predictable.

Usually, in nature, events that occur with a nonnegligible probability follow the Gaussian distribution. However, the "black swan" is far removed from this picture. Our intuition tells us mistakenly that "black swans" are even rarer than they are. Hence, we neglect their occurrence by moving them toward a much lower probability.

What can be the "black swan" in physics? Following our comments in subchapter 1.1 regarding vertical progress in classical physics, it was the discovery of the theory of relativity and quantum physics. In mathematics, the "black swans" (such as "Deus ex machina") were Gödel's incompleteness theorems and Lobachevskian geometry. Other rare and extremely rare events occur in many physical complex systems, mainly geological, meteorological, and climate events. Currently, the main task in the scientific community is to develop probabilistic and dynamical techniques to accurately estimate the probabilities of rare events, keeping in mind that it has to be done from limited data. Some of them include: (1) the generalized extreme value method from classical extreme value theory; (2) genealogical particle analysis and the Giardina-Kurchan-Lecomte-Tailleur algorithm; and (3) the brute force Monte Carlo [28].

There was a need for the new axiomatization of probability that required another treatment of the measurement of rare and frequent events on a deeper level. One such axiomatization was proposed by Chichilnisky [29], relying on the axiom of choice that lies at the very foundation of mathematics, where Gödel contributed significantly with his paper from 1940 [30]. In that paper, he proved that the axiom of choice and the axioms of von Neumann-Bernays-Gödel set theory are consistent. Chichilnisky's axiomatic approach to probability with the "black swan" includes three axioms about subjective probabilities related to: (1) continuity and additives; (2) unbiasing against rare events; and (3) unbiasing against frequent events. Those axioms imply that subjective probabilities are neither finitely additive nor countably additive but a combination of both. However, the finitely additive part allocates more weight to rare events than standard distributions.

No-event registered. It is possible that an event that could have happened *a priori* did not happen *a posteriori*. In this case of the event that is not registered, parameters cannot be estimated accurately. Still, a mathematical method for obtaining the boundaries of parameters is quite possible [1]. Zlokazov [31] offers different solutions to this problem, emphasizing the "absence of event" model. This method was applied to the data on the synthesis of the 114th element. Flerovium is an artificial chemical element with several short-living isotopes [32]. The signal was recorded over 34 minutes; the total experiment and the calibration time were about 48800 minutes each. All signals were recorded in the same strip detector. Calibration measurement resulted in the following frequencies: the implantation of recoil nuclei = 1.3 per hour; alpha particles = 1 per hour; in the case of spontaneous imitators, no division was recorded for the entire calibration time. An alternative hypothesis was that all signals were random imitators or temporary Poisson processes, perhaps the best formalism to describe random events. They are characterized by the number of random events occurring per constant time period, the memoryless waiting time between events (the independence from prehistory), and rarity (rare or low-probability events) [33]. The distribution function is

$$Q_k(t) = \frac{(lt)^k}{k!}\exp(-lt), \quad t \in [0,\infty], \qquad (5.4.1)$$

where l is the rate parameter of the Poisson distribution; t is the time. The quantity is the mathematical expectation and, at the same time, dispersion $k(t)$ at time t. For signals of different types, the formula describing the probabilities of their sums, taking into account their order and configuration, has the form

$$Q_{sk}(t) = p_s \prod_{i=1}^{m} Q_{ki}(t), t \in [0, \infty], \quad (5.4.2)$$

where m is the number of types, p_s is the probability of an ordered combination of s-type signals, and Q_{sk} is the total sum of these signals. Therefore, the mathematical expectation of the number of intervals having a similar configuration of signals over time (0.48800) was zero. It could be stated that the four signals contradict the hypothesis about their background origin [33].

5.5 Information in Complex Systems

What is information in complex systems? We can define it for some systems; however, we cannot find an exact answer to this question for many of them. We can only make a list of obstacles to the accessibility to information, which are, in our opinion, caused mainly by 1. interlocked interactions between the components of a complex system (i.e., one component of the system automatically triggers or prevents the action of another one); 2. The occurrence of additional information that originates from the system but is not visible to the observer. This hidden information is considered a hidden variable. If it is ignored, then an analysis of such a system is impossible. In other words, it is necessary to know how the parts of a complex system interact.

What is the situation regarding information in the physics of complex systems? Extracting information from many systems in nature is much more complicated than from systems in physics. However, physical systems have some characteristics that are not inherent to other systems. For systems that are described algorithmically as they are in physics, in general, we are provided with the information about 1. how the states of components evolve in time; 2. the evolution of its internal states (interactions) determining the update of the states at the next time step. The list that specifies physical complex systems, including the general characteristics of complex systems, can be summarized in the following manner: 1. Physical complex systems are often inherently nonergodic and non-Markovian. 2. The occurrence of the superpositions of interactions similar in strength. 3. Physical complex systems are often chaotic since they depend strongly on initial conditions and details of the system. 4. The equations of a physical complex system describing dynamics in an algorithmic manner are often nonlinear. 5. Physical complex systems are often driven systems. 6. Physical complex systems can exhibit emergence in a simple form, like the liquid phase of water, an emergent property of the atoms involved and their interaction [34, 35].

In general, methods for obtaining information from complex systems can be divided into methods for (1) analyzing data, (2) constructing and evaluating models, and (3) computing complexity. Here we focus only on algorithmic information theory. How can we measure the amount of information in the event given to us as an observable fact? Both Shannon information theory ("classical") and algorithmic information theory define it in terms of the minimum number of bits needed to describe the observation. However, unlike Shannon information theory, Kolmogorov's theory is not probabilistic: the amount of information is defined as the length of the *shortest* computer program [36] so that complexity and information are quantified in terms of the size of a program. Note that Kolmogorov dealt with complexity strongly motivated by information theory and problems associated with randomness.

Let us suppose that a universal Turing machine program generates the output $U(p) = x$. Then the Kolmogorov complexity $K(x)$ of a given object is defined as $K(x) = \min \{ \vee\, p \vee : U(p) = x\}$, which is the size of the shortest program p that generates the object x. $K(x) = x +$ const, while $K(x) \approx x$ for most objects. Since $K(x)$ of an arbitrary object x is incomputable, it is approximated by the Lempel-Ziv algorithm [37] (Appendix A) and its variants.

When compared with other methods for detecting regular behavior, particularly statistical methods and those from nonlinear dynamical systems, complexity has advantages because it does not assume stationary probabilities and the existence of a low-dimensional object in a phase space. $K(N)$ has different values for different types of time series: it is approximately 1. near zero for a periodic or regular time series; 2. halfway between zero and one for chaotic time series; and 3. near or larger than one for highly random time series [38]. The issue with Kolmogorov complexity is that it depends on the choice of the value used for the binarization of the series. This problem is addressed by the Kolmogorov complexity spectrum [39], in which complexity is calculated by taking each series element as a threshold. Ultimately, this relates the resulting probability distribution to compressibility. Although it is much more time-consuming than $K(x)$, the Kolmogorov complexity spectrum does not depend critically on the size of a time series; it does not need a vast amount of data to produce reliable values. Figure 5.2 depicts the Kolmogorov spectrum for different time series. It is easy to identify and differentiate the curves of the chaotic and quasi-periodic series from those of the random series. The random variations differ from each other, which reflects their probability distribution. The random series peak at the median value with respect to the probability distribution.

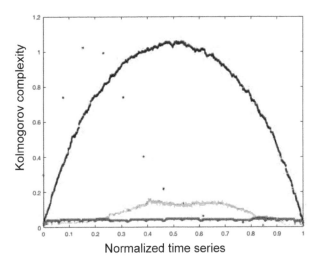

FIGURE 5.2
Kolmogorov complexity spectrum for different time series with sample sizes of 3000: (1) quasi-periodic (*the lower series*), (2) Lorenz attractor (*the series in the middle*), (3) random with the constant distribution (*the upper series*), and (4) random with the Poisson distribution (*dots*). (Figure courtesy of Marcelo Kovalsky.)

References

[1] D. T. Mihailović, A. Mihailović, C. Gualtieri, and D. Kapor, "How to assimilate hitherto inaccessible information in environmental sciences?," presented at 10th Int. Congr. on Environmental Modelling and Software, Brussels, Belgium, 2020.

[2] R. Landauer, "Information is a physical entity," *Physica A*, vol. 263, no. 1–4, pp. 63–67, Feb. 1999, doi: 10.1016/S0378-4371(98)00513-5.

[3] C. G. Timpson, "Philosophical aspects of quantum information theory," 2006, *arXiv*: quant-ph/0611187.

[4] M. J. Bates, "Fundamental forms of information," *J. Amer. Soc. Inf. Sci. Technol.*, vol. 57, no. 8, pp. 1033–1045, Apr. 2006, doi: 10.1002/asi.20369.

[5] V. Vedral, *Dekodiranje stvarnosti*, G. Skrobonja, Trans., Belgrade, Serbia: Laguna, 2014.

[6] R. V. L. Hartley, "Transmission of information," *Bell Syst. Tech. J.*, vol. 7, no. 3, pp. 535–563, Jul. 1928, doi: 10.1002/j.1538-7305.1928.tb01236.x.

[7] P. Adriaans, "Information," in *The Stanford Encyclopedia of Philosophy*, E. N. Zalta, Ed., Stanford University, 1997–. Available: https://plato.stanford.edu/entries/information/

[8] C. E. Shannon, "A mathematical theory of communication," *Bell Syst. Tech. J.*, vol. 27, no. 3, pp. 379–423, Jul. 1948, doi: 10.1002/j.1538-7305.1948.tb01338.x.

[9] A. Hájek, "Interpretations of probability,"in *The Stanford Encyclopedia of Philosophy*, E. N. Zalta, Ed., Stanford University, 1997–. Available: https://plato.stanford.edu/archives/fall2019/entries/probability-interpret/

[10] P. S. Laplace, *Philosophical Essay on Probabilities (Sources in the History of Mathematics and Physical Sciences 13)*, A. I. Dale, Trans., New York; Berlin, Germany; Vienna, Austria: Springer-Verlag, 1995.

[11] J. M. Keynes, *Treatise on Probability*. London, UK: Macmillan, 1921.

[12] C. Mungan and T. Lipscombe, "Probability analysis for rolls of a square cuboidal die," *Math. Gaz.*, vol. 97, no. 538, pp. 163–166, Mar. 2013, doi: 10.1017/S0025557200005635.

[13] A. Tarantola. (2004). Probability and information [PowerPoint slides]. Available: https://www.geophysik.uni-muenchen.de/~igel/downloads/inviiprob.pdf

[14] M. A. Nielsen and I. L. Chuang, *Quantum Computation and Quantum Information*. Cambridge, UK: Cambridge University Press, 2010.

[15] M. M. Vopson, "Estimation of the information contained in the visible matter of the universe," *AIP Adv.*, vol. 11, no. 10, Oct. 2021, Art. no. 105317, doi:10.1063/5.0064475.

[16] A. Ben-Naim, "Entropy, Shannon's measure of information and Boltzmann's H-theorem," *Entropy*, vol. 19, no. 2, p. 48, Jan. 2017, doi: 10.3390/e19020048.

[17] M. Tribus and E. C. McIrvine, "Energy and information," *Sci. Amer.*, vol. 225, no. 3, pp. 179–184, Sep. 1971, doi:10.1038/scientificamerican0971-179.

[18] J. V. Stone, "Information theory: A tutorial introduction," 2018, *arXiv*: 1802.05968.

[19] S. A. Umpleby, "Physical relationships among matter, energy and information," *Syst. Res. Behav. Sci.*, vol. 24, no. 3, pp. 369–372, May/Jun. 2007, doi:10.1002/sres.761.

[20] L. Szilard, "On the decrease of entropy in a thermodynamic system by the intervention of intelligent beings," (in German), *Z. Phys.*, vol. 53, pp. 840–856, Nov. 1929, doi: 10.1007/BF01341281.

[21] W. R. Ashby, "Some consequences of Bremermann's limit for information-processing systems," in *Cybernetic Problems in Bionics*, H. L. Oestreicher and D. R. Moore, Eds., New York, NY, USA: Gordon & Breach, 1968, pp. 69–79.

[22] H. J. Bremermann, "Optimization through evolution and re-combination," in *Self-Organizing Systems*, M. C. Yovits and G. T. Jacobi, Eds., Washington, DC, USA: Spartan, 1962, pp. 93–106.

[23] G. Gorelik, "Bremermann's limit and cGh-physics," 2009, *arXiv*: 0910.3424.

[24] S. Umpleby, "Some applications of cybernetics to social systems," Ph.D. dissertation, University of Illinois Urbana-Champaign, Urbana, IL, USA, 1975.

[25] M. M. Vopson, "The mass-energy-information equivalence principle," *AIP Adv.*, vol. 9, no. 9, Sep. 2019, Art. no. 095206, doi:10.1063/1.5123794.

[26] J. L. Beck and K. M. Zuev, "Rare-event simulation," in *Handbook of Uncertainty Quantification*, R. Ghanem, D. Higdon , and H. Owhadi, Eds., New York; Berlin, Germany; Vienna, Austria: Springer-Verlag, 2015, pp. 1–26.

[27] N. N. Taleb, *Crni labud*, A. Imširović Đorđević and A. Ješić, Trans., Smederevo, Serbia: Heliks, 2016.

[28] M. Carney, H. Kantz, and M. Nicol, "Analysis and simulation of extremes and rare events in complex systems," in *Advances in Dynamics, Optimization and Computation* (Studies in Systems, Decision and Control 304), O. Junge, O. Schütze, G. Froyland, S. Ober-Blöbaum, and K. Padberg-Gehle, Eds., New York; Berlin, Germany; Vienna, Austria: Springer-Verlag, 2020, pp. 151–182.

[29] G. Chichilnisky, "The foundations of probability with black swans," *J. Probab. Stat.*, vol. 2010, Mar. 2010, Art. no. 838240, doi: 10.1155/2010/838240.
[30] K. Gödel, *Consistency of the Continuum Hypothesis* (Annals of Mathematics Studies 3). Princeton, NJ, USA: Princeton University Press, 1940.
[31] V. B. Zlokazov, "No event registered—what information can be derived from this?" (in Russian), Joint Inst. Nucl. Res., Dubna, Russia, Rep. JINR-2-117-2003, Jul. 2003.
[32] Y. U. T. S. Oganessian et al., "Synthesis of superheavy nuclei in the reactions of ^{244}Pu and ^{248}Cm with ^{48}Ca," in *Proc. Int. Symp.*, Lake Baikal, Russia, Jul. 2001, pp. 21–33, doi: 10.1142/9789812777300_0002.
[33] V. B. Zlokazov, "Statistical analysis of rare events—synthesis of the element 114," *Eur. Phys. J. A*, vol. 8, no. 1, pp. 81–86, Jul. 2000, doi: 10.1007/s100500070121.
[34] R. Hanel and S. Thurner, "Generalized (c,d)-entropy and aging random walks," *Entropy*, vol. 15, no. 12, pp. 5324–5337, Dec. 2013, doi: 10.3390/e15125324.
[35] S. Thurner, R. Hanel, and P. Klimek, *Introduction to the Theory of Complex Systems*, London, UK: Oxford University Press, 2018.
[36] P. Harremoes and F. Topsoe, "Inequalities between entropy and index of coincidence derived from information diagrams," *IEEE Trans. Inf. Theory*, vol. 47, no. 7, pp. 2944–2960, Nov. 2001, doi: 10.1109/18.959272.
[37] A. Lempel and J. Ziv, "On the complexity of finite sequence," *IEEE Trans. Inf. Theory*, vol. 22, no. 1, pp. 75–81, Jan. 1976, doi: 10.1109/TIT.1976.1055501.
[38] M. G. Kovalsky, A. A. Hnilo, and M. B. Agüero, "Kolmogorov complexity of sequences of random numbers generated in Bell's experiments," *Phys. Rev. A*, vol. 98, no. 4, Oct. 2018, Art. no. 042131, doi: 10.1103/PhysRevA.98.042131.
[39] D. T. Mihailović, G. Mimić, E. Nikolić-Djorić, and I. Arsenić, "Novel measures based on the Kolmogorov complexity for use in complex system behavior studies and time series analysis," *Open Phys.*, vol. 13, no. 1, pp. 1–14, 2015, doi: 10.1515/phys-2015-0001.

6

Kolmogorov and Change Complexity and Their Applications to Physical Complex Systems

6.1 Kolmogorov Complexity: An Incomputable Measure and Lempel-Ziv Algorithm

One of the most fundamental concepts in algorithmic information theory is that of Kolmogorov complexity ($K(x)$) [1]. The motivation behind Kolmogorov complexity was well described by Grunwald and Vitanyi [2]:

> Suppose we want to describe a given object by a finite binary string. We do not care whether the object has many descriptions; however, each description should describe but one object. From among all descriptions of an object we can take the length of the shortest description as a measure of the object's complexity. It is natural to call an object 'simple' if it has at least one short description, and to call it 'complex' if all of its descriptions are long.

The Kolmogorov complexity of an object represented by a finite binary sequence is defined as the minimum length of the binary sequence needed to reconstruct the whole object. Consequently, the object is more complex if it is described with a longer sequence and vice versa. It was shown that Kolmogorov complexity is independent of a programming language used to compute it up to the constant.

The well-known limitation of Kolmogorov complexity is that it is generally incomputable and can only be approximated. For practical applications, Kolmogorov complexity can be approximated by using some data compressor $C(x)$; furthermore, $C(x) \approx K(x)$ for most sequences [2]. The most famous algorithm (LZA) [3] was developed in 1976 and was improved [4] in 1984 (the LZW algorithm; Appendix A). LZA counts the minimal number of distinct patterns in a given time series [5]. The additional problem with LZA is that, although it is commonly applied to longer sequences in different areas (such as physics, biology, or neurology), it is inapplicable to short sequences. Change complexity [6] is more suitable for the complexity of short sequences. Also, an interesting no-threshold encoding scheme for calculating the Kolmogorov complexity of short sequences was suggested in [7] in an analysis of day-to-day hourly solar radiation time series.

We list some technical details about the applications of LZA, which come from our experience in numerical and laboratory experiments with the logistic equation and velocity time series of fluid flow (in the turbulent regime) [8]. *The influence of increasing the numbers of samples in a time series.* For the logistic equation with the initial condition $x_0 = 0.2$, the logistic parameter $r = 3.8$, and the iterative step 1, the size of the time series was expanded up to 270,000. The Kolmogorov complexity of the generated time series decreased exponentially because the number of distinct patterns (i.e., the complexity counter $c(N)$) had the saturation value caused by the increase of the denominator $b(N)$ (for the explanation of the complexity counter and denominator see Appendix A). Interestingly, the complexity counter detected 301 patterns in 6112 iterations, while only two more new patterns were found in the remaining iterations. These results indicate that long-term behavior described with the logistic map is not entirely random and that some "hidden" rules govern the dynamics of the logistic map. In the time series with the fluid flow velocity in the turbulent regime (270,000 samples), Kolmogorov complexity increased up to 0.6 when the size was around 85,000 and then maintained that value.

The Running Kolmogorov Complexity. In the physics of complex systems, the running complexity of a time series is calculated by applying the complexity algorithm at each point obtained by moving forward a window of fixed size by a one-time step [6]. The choice of the window size depends on the nature of the time series. The running Kolmogorov complexity helps us to obtain information from the time series about how a complex physical system behaves in a longer period and in which mode are interactions in that system. We take the logistic equation as an example since it was already shown (see subchapters 3.3 and 4.5) that it describes how radiation interacts with the surface as well as the response of that surface. The selected values of the logistic parameters are such that the system is highly chaotic in both modes. Figure 6.1 illustrates potential modes in which a complex system can be due to the interactions of its components and the surrounding environment. In one mode, the system remains in a state of high randomness ($r = 3.951$, where r is the parameter of the logistic equation). In another mode ($r = 3.950$), the system is characterized by high complexity with one "break" that begins to decline following the power law. In other words, from a point corresponding to a window size of about 1200 scaling time units, one microscopic process becomes "visible" and maybe macroscopically measurable.

We call this point the "breaking" point as an analogy with the inflection point in mathematics. However, this analogy is incomplete since the time

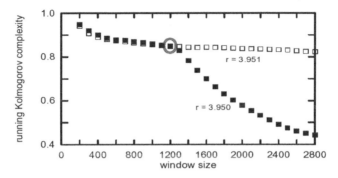

FIGURE 6.1
The running Kolmogorov complexity of time series of the logistic equation for different fixed moving windows (200, 300, 400, ..., 2800). Calculations were performed for two time series, each of length 3000, with values of the logistic parameter $r = 3.950$ and $r = 3.951$. The black squares correspond to the values obtained for $r = 3.950$, while the circle is the "breaking" point.

evolution of complexity is represented in a discrete form. This behavior of a physical complex system often occurs during the interaction of gamma rays with some environmental interface or during their detection [9]. Note that the system's complexity takes a constant value (either high or low complexity) after a long time, irrespective of whether a time series is obtained by model or measurements. Finally, let us mention the relationship between chaos and complexity and an illustrative comment by Cristoforo Bertuglia and Franco Vaio: "Stand alone chaos and complexity have absolutely nothing to do with generating formal function. Chaotic systems are not necessarily complex, and complex systems are not necessarily chaotic" [10].

6.2 Change Complexity: A Measure that Detects Change

Shannon information theory and algorithmic information theory have been the dominant approaches to the complexity and randomness of complex systems. Aksentijevic and Gibson [11] introduce a new approach and measure change complexity (AG complexity) based on the amount of change incorporated in a binary sequence. The theoretical background and detailed descriptions of AG complexity are given in [6, 11], while their mutual summary and mathematically rigorous extension are given in Appendix B. Let $S = (s_1, s_2, ..., s_L)$ be a binary sequence of the length L. The *change* in S is defined in the following manner: $S = (s \mid 1, s_2)$ has change if $s_1 \neq s_2$; $S = (s \mid 1, s_2, s_3)$ has change if $(s \mid 1, s_2)$ has change and $(s \mid 2, s_3)$ does not or vice versa; in general, $S = (s_1, s_2, ..., s_L)$, $L > 3$, has change if $(s_1, s_2, ..., s_{L-1})$ has change and $(s_2, s_3, ..., s_L)$ does not or vice versa. The number of changes p_i registered for each subsequence of S forms the profile $P = (p_2, p_3, \ldots, p_L)$. The change complexity C of S is

$$C = \sum_{i=2}^{L} p_i w_i, \quad w_i = \frac{1}{L-i+1}, \qquad (6.2.1)$$

while normalized complexity N of S is

$$N = \frac{C}{L-1}. \qquad (6.2.2)$$

The values of C and N lie in the ranges $0 \leq C < L - 1$, $0 \leq N < 1$, and N_{min}, $N_{max} \to 1$ as L gets large. Both profile and complexity indicate the amount of change. Thus, the registered changes of each subsequence contribute to the overall complexity of S.

AG complexity is computable and can be applied to sequences of any length, contrary to Kolmogorov complexity, which is incomputable and inapplicable to short sequences (see the previous subchapter). One of the characteristics of change complexity is that it can detect *periodicity* in a time series that is often undetectable regardless of the information measure or physical method used for analysis. AG complexity correlates well with the measures of complexity based on periodicity, self-similarity, and symmetry, representing the first step toward merging "subjective" and "objective" perspectives on complexity in the analysis of binary time series in the physics of complex systems. AG complexity was also implemented for 2D arrays allowing the investigation of the spatial arrangements of various types of data—from dissipative processes to magnetic resonance imaging scans (an algorithm for its computation is provided in several papers [6, 11–12]).

We want to demonstrate how information at a small scale can be extracted from a complex physical phenomenon—that is, whether irregularities in a physical time series can be detected by analyzing its short segments with the running AG complexity. Fluctuations in sea surface temperatures cause the El Niño-Southern Oscillation (ENSO). The warming and cooling of the ocean surface in the central and eastern tropical Pacific Ocean are El Niño and La Niña, respectively. To classify and analyze ENSO events, we used ENSO episodes categorized by the US National Oceanic and Atmospheric Administration (NOAA) [13] and its data set [14] that consists of two series of sea surface temperatures: a series with temperature anomalies and a reference one. A drop in the running AG complexity, especially one whose value is zero (zero-complexity drop), can indicate disturbances in a time series and affected segments. The left panel of Figure 6.2 shows that no such complexity drops were observed in the reference series, while many occurred in the other series; therefore, the temperature anomalies were identified correctly

Kolmogorov and Change Complexity and Their Applications 85

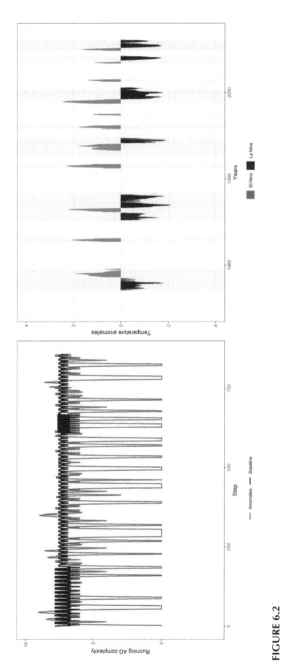

FIGURE 6.2
Left, the running AG complexities of the two sea surface temperatures for the window size 12. *Right*, the regions of low complexity (*rectangles*) paired up with El Niño (*the area plot above zero*) and La Niña (*the area plot below zero*) events from 1954 to 2012.

by change complexity. These regions of lower complexity correspond to El Niño and La Niña events from 1954 to 2012 (the right panel of Figure 6.2). Additional regions of low complexity coincide with the temperature anomalies that are not, according to their strength, classified neither as El Niño nor as La Niña.

AG complexity can also be applied to strings. Contrary to complexity values, their distribution depends on the size of the alphabet. Aksentijevic and Gibson [15] use the measure and distributions of complexity values of substrings of a string to create *a structure surface*. The structure surface is constructed by plotting the percentiles of complexity distributions of substrings against the running complexity (offset) and length of a substring (window size) (Figure 6.3). To get distribution for the substring of the length i, one million random strings of the length i were generated.

A pocket is the region of the structure surface containing values lower than approximately 10–15 percentiles or a region of low complexity. If the pocket is distributed over a wide range of window sizes, it is called a large pocket. There are also many small regions of low complexity; however, they disappear as length size increases. The bottom edge of the pocket is termed the width of the pocket. The substring whose first element is located at the start of the width of the pocket and is twice larger than the width of the pocket is a critical segment. This string has low complexity, and if it is substituted with a high-complexity string, the pocket disappears. It was observed that the minor modifications of strings with low complexity result in strings with high complexity, which means that small changes to the critical segment significantly change its structure surface. Structure surfaces whose critical segments are strings with complexity 1 are especially sensitive since modifying

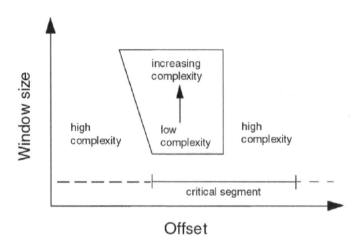

FIGURE 6.3
The structure surface of a string. (Reproduced by permission from [15].)

only 15–20% of their letters causes their complexity percentiles to increase to over 90%.

Structure surfaces allow us to visualize the structure of longer strings and might be valuable indicators in DNA analysis. It must be emphasized that the characteristics of structure surfaces mentioned above are observed in both DNA sequences and random strings. Some preliminary empirical results confirmed the existence of a large pocket in a mutated cancer human DNA sequence. Additionally, the structure surface of human rhodopsin differed from the structure surfaces of pig and mouse rhodopsin.

6.3 Kolmogorov Complexity in the Analysis of the LIGO Signals and Bell's Experiments

In order to demonstrate the broad range of physical problems where using Kolmogorov complexity might provide a substantial contribution, here we present how KC was applied in analyzing the data from the class of problems that lead to two recent Nobel Prizes in Physics. 1. In 2017 Rainer Weiss, Barry Barish, and Kip Thorne were awarded "for decisive contributions to the LIGO detector and the observation of gravitational waves" [16]. 2. In 2022 Alain Aspect, John Clauser, and Anton Zeilinger were awarded "for experiments with entangled photons, establishing the violation of Bell inequalities and pioneering quantum information science" [17].

Kolmogorov complexity spectrum of the raw, template, and residual time series in LIGO experiments. On September 14, 2015, it was announced that gravitational waves were detected. From that moment, the LIGO (Laser Interference Gravitational-Waves Observatory) story has been accompanied by dilemmas that raise new questions and confusion related to this detector and institutions participating in the LIGO experiment. Issues include scientific, engineering, technological, and social aspects. On February 11, 2016, LIGO announced the detection of the gravitational wave event GW150914 [18]. Four months later, they detected another two gravitational events, WG151226 and LVT151012 [19]. The spokesperson for the LIGO Scientific Collaboration says,

> The physics is really the easy part. More complex was the delicate precision-engineering in designing, upgrading and now maintaining the facilities needed to detect black hole collisions more than a billion light years away. Add to that the more earthly challenges of getting a group of about 1,000 opinionated scientists in a dozen countries to work on a common project, interpret complex data, and agree on wording for a research paper.
>
> [20]

In our opinion, this is the most accurate description of a significant endeavor in physics that 1. is the technologically impressive and expensive project; 2. offers a lot of room for the discussion and publication of papers in the most prestigious journals; 3. leaves a space for skepticism partly elaborated—for example, in [21]; 4. is suitable for the more subtle decoding of information from its outcomes. Different aspects of this experiment are discussed in subchapter 1.2.

Vilasi's paper [22] is a good guide for understanding gravitational waves as the exact solutions of Einstein's field equations. In this mathematically elegant paper, it is 1. described that these exact solutions of Einstein's field equations are invariant for a non-Abelian two-dimensional Lie algebra of Killing fields; 2. shown that a sub-class of these gravitational fields has a wave-like character with spin−1 [23]. The elegance and persuasiveness of Vilasi's paper lead us to think that gravitational waves, as the exact solution of Einstein's field equations, do not even need experimental proof. This idea is popular among many physicists. Albert Einstein anticipated that gigantic objects, such as black holes and neutron stars, would send out ripples overspreading in space-time during their collision. Interestingly, Einstein did not believe humans would ever detect these ripples or gravitational waves. Comments on the possibilities of detecting gravitational waves by LIGO are various. They range from the objection that the basic principle of the LIGO experiment is wrong—that is, LIGO experiments cannot detect gravitational waves by using laser Michelson interferometers [24]—to technical characteristics, since LIGO is principally insensitive in the frequency range of 10^{-4} Hz to 1 Hz, in which strong signals are expected from black hole formation.

The interpretation of the LIGO time series is an important issue. Any contribution to the interpretation of GW (gravitational waves) time series can be a small step toward a better understanding of information. Analyzing the time series recorded in the event GW150914 (Figure 6.4), Kovalsky and Hnilo [25] propose Takens's dimension of embedding and the Kolmogorov complexity spectrum to help in the identification of signals of cosmological interest.

At the core of the method for identifying GW is a comparison of the raw time series recorded by interferometers with a set of templates. These templates correspond to signals produced by plausible cosmological sources and are calculated according to the theory of general relativity. In the further procedure of the LIGO data analysis, the templates are subtracted from the raw signals to form the residual series, which should not result in the serial correlation of the residual series. If this is the case, then it casts methodological doubts about the origin of the observed signals; however, entirely uncorrelated elements of the residual series have not been noticed thus far. In the LIGO data analysis, information measures may remarkably provide additional information that is not provided by conventional methods. In other words, this additional information adds more certainty to an uncertain

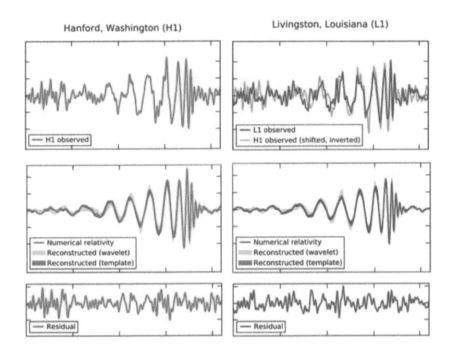

FIGURE 6.4
The recorded series of the event GW150914 at Hanford (*the left column*) and Livingston (*the right column*). The upper row corresponds to the two raw series, the middle row to the theoretical templates, and the row at the bottom to what is left after subtracting the templates from the raw series. (Reproduced by permission from [25].)

path to reaching the goal. Such a tool is the Kolmogorov complexity spectrum [26], in which complexity is calculated by taking each series element as a threshold. Accordingly, it does not depend on the choice of the threshold and also evaluates the maximum possible complexity (see subchapter 5.5 and Appendix A). In addition, total complexity is defined by the area under the Kolmogorov complexity curve. This derivate of Kolmogorov complexity (KC) is especially suitable for analyzing the series recorded in LIGO experiments since it is not influenced by the correlation between the residual series [25]. Figure 6.5 shows the Kolmogorov complexity spectra for the residual, raw, and template (generated by an algorithm) time series. It can be seen that KC for the templates is much smaller than KC for the residual noises. This is usually because noise unavoidably includes a degree of uncertainty. The raw time series has an intermediate value of KC.

Kolmogorov complexity in the analysis of Bell's experiments. Quantum entanglement, assumed (but also opposed) by Albert Einstein, has passed stringent tests gradually. However, doubts were not unusual if we consider that the

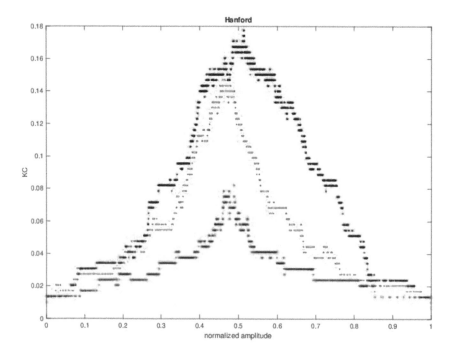

FIGURE 6.5
The Kolmogorov complexity spectrum of the series recorded at Hanford: residual (*the upper series*), raw (*the series in the middle*), and template (*the lower series*). The curves of the series recorded at Livingston are similar. (Reproduced by permission from [25].)

well-known quotation "Act locally, think globally!" can also be rearranged as "Act quantum mechanically, think cosmologically!" Einstein *et al.* [27] argue that the existence of the "elements of reality" is not a part of quantum theory. They also 1. speculate that it should be possible to construct a theory containing them (hidden variable theory); 2. state that if quantum mechanics fully describes reality, then making measurements of one part of an entangled system will instantaneously affect our knowledge about future measurements of the other part. Apparently, it implies that this information is sent at a speed higher than the speed of light (think cosmologically).

Later it became more evident that "spooky action at a distance" just seemed "spooky." What makes us think like that? There are several reasons: 1. The world of quantum mechanics, whose physics governs the universe's behavior at the smallest scales, is often described as outstandingly peculiar. It is strange to our experience and intuition (see subchapter 1.2). 2. According to the laws of quantum mechanics, nature's building blocks are waves and particles. It means that there is no possibility of "seeing" the definite location of a particle in space. 3. Measuring and observing a system pushes them to "choose" a definite state; maybe the quantum mechanics world has its own choice that is hidden from us because of "tyranny" laws that we imposed (see subchapter 4.1).

Although Bell's research has thrown light on the further study, it has also raised doubts about the approach to quantum mechanics. Because of the conciseness in the description of his work, we cite the text from [28].

> *Bell's Theorem* is the collective name for a family of results, all of which involve the derivation, from a condition on probability distributions inspired by considerations of local causality, together with auxiliary assumptions usually thought of as mild side-assumptions, of probabilistic predictions about the results of spatially separated experiments that conflict, for appropriate choices of quantum states and experiments, with quantum mechanical predictions. These probabilistic predictions take the form of inequalities that must be satisfied by correlations derived from any theory satisfying the conditions of the proof, but which are violated, under certain circumstances, by correlations calculated from quantum mechanics. Inequalities of this type are known as *Bell inequalities*, or, sometimes, *Bell-type inequalities*. Bell's theorem shows that no theory that satisfies the conditions imposed can reproduce the probabilistic predictions of quantum mechanics under all circumstances.

Entanglement itself has been verified over numerous decades, but even though we know that it is the experimental reality, we still need a convincing story about how it works.

A sequence of bits formed by tossing a quantum coin is genuinely random [29]. This result can be interpreted as the strengthening of Bell's no-hidden-variables theorem. When we think about randomness, it encompasses two notions—pseudo-randomness and algorithmic randomness. A pseudo-random sequence of numbers appears statistically random, although it can be generated by a completely deterministic and repeatable process. Statistical randomness is defined by the absence of regularities in a sequence, and algorithmic randomness is described in terms of the compressibility of a sequence. Surprisingly, little or no attention has been given to algorithmic randomness in analyzing the time series generated in Bell's experiments. Why is Kolmogorov complexity necessary to analyze outcomes in these experiments? A possibly very concise answer is given by Kovalsky *et al.* [30] in their paper analyzing sequences generated by the entangled pairs of photons for the main set of the data of Bell's experiment performed in Innsbruck [31] and the data recorded in their setup.

> In our opinion, the main conclusion of this study is that, although random sequences are generated in most cases, it is not safe taking randomness for granted in experimentally generated sequences, even if they violate the related Bell's inequality by a wide margin with a maximally entangled state. Deviations from randomness are observed even in the controlled conditions of the Innsbruck experiment, which are very difficult (perhaps impossible) to achieve in a QKD setup operating in a real-

world situation. Therefore, applying additional statistical and algorithmic tests and, if necessary, using distillation and extraction techniques are advisable before coding a message in the practice [30].

6.4 Change Complexity in the Search for Patterns in River Flows

The behavior of environmental fluids is influenced by human-induced activities, climate change, and increasing water pollution. As a result, streamflow varies with respect to time and space and exhibits different levels of complexity—from simple to complex. Rivers are the part of complex natural fluid systems; therefore, the behavior of their constituents may be beyond our understanding. Information about those complex systems is only available if it is obtained by analyzing streamflow time series. Entropy is often used to analyze those time series (including different forms of entropy and heuristic models), but its practical solutions are not always optimal [32]. Kolmogorov complexity and its different forms (an average measure, spectrum, and running mean) may give more useful intrinsic information than entropy measures [33]. However, they are rarely used in the hydrological analysis. The analysis of streamflow (or any natural time series) with change complexity may reveal periodicity that can be attributed to specific environmental factors and human activity [34]. Because of the complex nature of environmental fluids, finding patterns in streamflow time series that may be regular is of great importance. Related to this remark, we characterize the following points: 1. Regularity is a repetition of one or more conditions at equal intervals. 2. If periodicity is detected, it may not be preserved over a longer time period [34].

We demonstrate the change complexity analysis [35] of the ten streamflow time series (Figure 6.6) from the seven rivers in Bosnia and Herzegovina for the period 1965–1986 at different temporal scales.

General analysis. The running change complexities of the rivers (the window size $L = 128$) were compared with the running complexity of a random sequence to discriminate between regular and random behavior. A stationary pattern was present in the flow of all rivers and contained the approximately annual period with bimodal peaks. Most rivers were strongly correlated, confirming the existence of the similar rule that rivers followed irrespective of a geographical position and distance. The only river that was not associated with others was Ukrina (UKR_D) since it has little connection with the dynamics of the high-altitude hydro-system to the South. River complexities were much lower than the complexity of the random sequence,

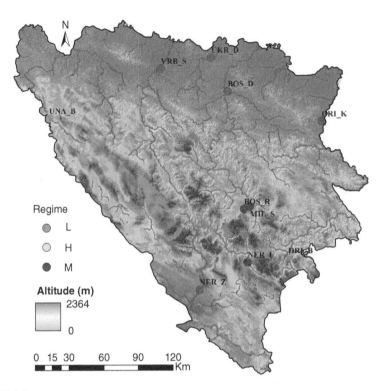

FIGURE 6.6
The relief of Bosnia and Herzegovina with the locations of ten hydrological stations on seven rivers. The rivers are classified as follows: (1) lowlands (0–200 m mean altitude – L type), (2) platforms and hills (200–500 m – H type), and (3) mountains with mean elevations between 500 and 6000 m (M type). (Reproduced by permission from [36].)

which was in agreement with the presence of the pattern in the time series. At this scale, change complexity also detected a decreasing trend—that is, long-term effects that reduce the complexity of river flow. Complexity values were compatible with the hypothesis that mountainous rivers are expected to have higher complexity.

Temporal analysis at a smaller scale. For the window size $L = 8$, zero-complexity drops were observed in each time series; however, their number and position varied across the rivers. The river Neretva to Ulog (NER_U) had the highest complexity, followed by two drops, while the river Bosna to Doboj (BOS_D) had nine such drops. This diversity was associated with a possible impact of precipitation on complexity, which was more prominent for the slower rivers. In line with the overall decrease in the complexity of river flow over time, most zero-complexity drops were located in the second part of the time series. Finally, the river Ukrina, which is not the part of the high-altitude rivers, had several drops ("phase-shifted" drops) that did not belong to the general annual behavior of other rivers. The most crucial finding identified by change complexity was periodicity with a period of 12 months. According to the information available to us, such regularity in river flow has not been detected until now and opens up new possibilities for predictability in hydrology. It should be stressed that the choice of the window size affects the result—that is, periodicity or other regularities may not be detected if the size of the window is changed, which may indicate time periods that are important or not important for the dynamics of river flow.

Structural analysis. This analysis includes examining time series by using more windows instead of just one and creating a time series structure map. The running change complexity values of the rivers were analyzed across various temporal scales. The river with the highest complexity lacked clear periodicity, and zero drops were small. Periodicity was similarly not observed in the river with the second highest complexity value; however, its zero-complexity drops were larger, which resulted in the lower complexity value. Therefore, the two rivers with the highest complexity values followed a similar rule characterized by the absence of periodicity and relatively small zero drops. These two rivers are mountainous; accordingly, they were expected to exhibit higher complexity and less regularity. On the other hand, the rivers with the lowest complexity values were characterized by periodicity at some time scales. These high- and low-complexity cycles were ascribed to changes in precipitation and the melting of snow that increase and decrease river levels [35].

References

[1] A. N. Kolmogorov, "Three approaches to the quantitative definition of information," (in Russian), *Probl. Peredachi Inf.*, vol. 1, no. 1, pp. 3–11, 1965.
[2] P. D. Grunwald and P. M. B. Vitanyi, "Algorithmic information theory," 2008, *arXiv*: 0809.2754.
[3] A. Lempel and J. Ziv, "On the complexity of finite sequences," *IEEE Trans. Inf. Theory*, vol. 22, no. 1, pp. 75–81, Jan. 1976, doi: 10.1109/TIT.1976.1055501.
[4] T. A. Welch, "A technique for high-performance data compression," *IEEE Trans. Comput.*, vol. 17, no. 6, pp. 8–19, Jun. 1984, doi: 10.1109/MC.1984.1659158.
[5] F. Kaspar and H. G. Schuster, "Easily calculable measure for the complexity of spatiotemporal patterns," *Phys. Rev. A*, vol. 36, no. 2, pp. 842–848, Jul. 1987, doi: 10.1103/PhysRevA.36.842.
[6] A. Aksentijevic, D. T. Mihailović, D. Kapor, S. Crvenković, E. Nikolić-Djorić, and A. Mihailović, "Complementarity of information obtained by Kolmogorov and Aksentijevic–Gibson complexities in the analysis of binary time series," *Chaos Solitons Fractals*, vol. 130, Jan. 2020, Art. no. 109394, doi: 10.1016/j.chaos.2019.109394.

[7] M. Bessafi, D. T. Mihailović, P. Li, A. Mihailović, and J.-P. Chabriat, "Algorithmic probability method versus Kolmogorov complexity with no-threshold encoding scheme for short time series: An analysis of day-to-day hourly solar radiation time series over tropical western Indian Ocean," *Entropy*, vol. 21, no. 6, p. 552, May 2019, doi: 10.3390/e21060552.

[8] G. Mimić, I. Arsenić, and D. T. Mihailović. Number of patterns in Lempel-Ziv algorithm applied to iterative maps and measured time series. Presented at the *8th Int. Congr. Environ. Model. Softw.*, Toulouse, France. Jul. 2016, [Online]. Available: https://scholarsarchive.byu.edu/iemssconference/2016/Stream-A/63/

[9] D. T. Mihailović, S. Avdić, and A. Mihailović, "Complexity measures and occurrence of the "breakpoint" in the neutron and gamma-rays time series measured with organic scintillators," *Radiat. Phys. Chem.*, vol. 184, Jul. 2021, Art. no. 109482, doi:10.1016/j.radphyschem.2021.109482.

[10] C. S. Bertuglia and F. Vaio, *Nonlinearity, Chaos, and Complexity: The Dynamics of Natural and Social Systems*. London, U.K.: Oxford University Press, 2005.

[11] A. Aksentijevic and K. Gibson, "Complexity equals change," *Cogn. Syst. Res.*, vol. 15–16, pp. 1–16, May 2012, doi:10.1016/j.cogsys.2011.01.002.

[12] A. Aksentijevic, A. Mihailović, and D. T. Mihailović, "Time for change: Implementation of Aksentijevic-Gibson complexity in psychology," *Symmetry*, vol. 12, no. 6, p. 948, Jun. 2020, doi: 10.3390/sym12060948.

[13] "Past events." *Noaa.gov*. https://psl.noaa.gov/enso/past_events.html (accessed Apr. 1, 2022).

[14] *Detrend.nino34.ascii*, National Oceanic and Atmospheric Administration. [Online]. Available: https://www.cpc.ncep.noaa.gov/products/analysis_monitoring/ensostuff/detrend.nino34.ascii.tx

[15] A. Aksentijevic and K. Gibson, "Structure surfaces of strings and DNA sequences," unpublished.

[16] "The Nobel Prize in Physics 2017." *Nobelprize.org*. https://www.nobelprize.org/prizes/physics/2017/summary/ (accessed 20 Dec. 2022).

[17] "The Nobel Prize in Physics 2022." *Nobelprize.org*. https://www.nobelprize.org/prizes/physics/2022/summary/ (accessed 20 Dec. 2022).

[18] B. P. Abbott et al., "Observation of gravitational waves from a binary black hole merger," *Phys. Rev. Lett.*, vol. 116, no. 6. Feb. 2016, Art. no. 061102, doi: 10.1103/PhysRevLett.116.061102.

[19] B. P. Abbott et al., "GW151226: Observation of gravitational waves from a 22-solar-mass binary black hole coalescence," *Phys. Rev. Lett.*, vol. 116, no. 24, Jun. 2016, Art. no. 241103, doi: 10.1103/PhysRevLett.116.241103.

[20] K. Bourzac, "Proving Einstein right," *Nature*, vol. 551, no. 7678, pp. S21–S23, Nov. 2017, doi: 10.1038/551S21a.

[21] D. Castelvecchi, "Has giant LIGO experiment seen gravitational waves?," *Nature*, Sep. 2015, doi: 10.1038/nature.2015.18449.

[22] G. Vilasi, "Gravitational waves as exact solutions of Einstein field equations," *J. Phys. Conf. Ser.*, vol. 87, Dec. 2007, doi: 10.1088/1742-6596/87/1/012017.

[23] F. Canfora, L. Parisi, and G. Vilasi, "Spin-1 gravitational waves: Theoretical and experimental aspects," *Int. J. Geom. Methods Mod. Phys.*, vol. 3, no. 3, pp. 451–469, May 2006, doi: 10.1142/S0219887806001247.

[24] X. Mei, Z. Huang, P. Yŏshin Ulianov, and P. Yu, "LIGO experiments cannot detect gravitational waves by using laser Michelson interferometers—light's wavelength and speed change simultaneously when gravitational waves exist which make the detections of gravitational waves impossible for LIGO experiments," *J. Mod. Phys.*, vol. 7, no. 13, pp. 1749–1761, Sep. 2016, doi: 10.4236/jmp.2016.713157.

[25] M. G. Kovalsky and A. A. Hnilo, "LIGO series, dimension of embedding and Kolmogorov's complexity," *Astron. Comput.*, vol. 35, Apr. 2021, Art. no. 100465, doi:10.1016/j.ascom.2021.100465.

[26] D. T. Mihailović, G. Mimić, E. Nikolić-Djorić, and I. Arsenić, "Novel measures based on the Kolmogorov complexity for use in complex system behavior studies and time series analysis," *Open Phys.*, vol. 13, no. 1, pp. 1–14, 2015, doi: 10.1515/phys-2015-0001.

[27] A. Einstein, B. Podolsky, and N. Rosen, "Can quantum-mechanical description of physical reality be considered complete?," *Phys. Rev.*, vol. 47, no. 10, pp. 777–780, May 1935, doi: 10.1103/PhysRev.47.777.

[28] Myrvold, Wayne, M. Genovese, and A. Shimony. "Bell's Theorem." In *The Stanford Encyclopedia of Philosophy*, E. N. Zalta, Ed. Stanford University Press, 1997–. Available: https://plato.stanford.edu/archives/fall2021/entries/bell-theorem/

[29] D. Abergel, "A quantum coin toss," *Nat. Phys.*, vol. 14, p. 7, Jan. 2018, doi: 10.1038/nphys4342.

[30] M. G. Kovalsky, A. A. Hnilo, and M. B. Agüero, "Kolmogorov complexity of sequences of random numbers generated in Bell's experiments," *Phys. Rev. A*, vol. 98, no. 4, Oct. 2018, Art. no. 042131, doi: 10.1103/PhysRevA.98.042131.

[31] G. Weihs, T. Jennewein, C. Simon, H. Weinfurter, and A. Zeilinger, "Violation of Bell's inequality under strict Einstein locality conditions," *Phys. Rev. Lett.*, vol. 81, no. 23, pp. 5039–5043, Dec. 1998, doi:10.1103/PhysRevLett.81.5039.

[32] D. T. Mihailović, I. Balaž, and D. Kapor, *Time and Methods in Environmental Interfaces Modelling: Personal Insights* (Developments in Environmental Modelling 29). Amsterdam, The Netherlands; New York, NY, USA: Elsevier, 2016.

[33] D. T. Mihailović, E. Nikolić-Đorić, N. Drešković, and G. Mimić, "Complexity analysis of the turbulent environmental fluid flow time series," *Physica A*, vol. 395, pp. 96–104, Feb. 2014, doi: 10.1016/j.physa.2013.09.062.

[34] D. T. Mihailović, A. Aksentijevic, and A. Mihailović, "Mapping regularities in the solar irradiance data using complementary complexity measures," *Stoch. Environ. Res. Risk Assess.*, vol. 35, no. 6, pp. 1257–1272, Jun. 2021, doi: 10.1007/s00477-020-01955-1.

[35] A. Aksentijevic, D. T. Mihailović, A. Mihailović, and V. P. Singh, "Regime-related regularities in river flow revealed by Aksentijevic-Gibson complexity," *J. Hydrol.*, vol. 598, Jul. 2021, Art. no. 126364, doi: 10.1016/j.jhydrol.2021.126364.

[36] D. T. Mihailović, G. Mimić, N. Drešković, and I. Arsenić, "Kolmogorov complexity based information measures applied to the analysis of different river flow regimes," *Entropy*, vol. 17, no. 5, pp. 2973–2987, May 2015, doi:10.3390/e17052973.

7

The Separation of Scales in Complex Systems: "Breaking" Point at the Time Scale

7.1 The Generalization of Scaling in Gödel's World. Scaling in Phase Transitions and Critical Phenomena

Common expressions describing physical phenomena at a macroscopic scale are no longer valid in the microscopic world governed by different laws. When we say "scale" in regular communication between physicists, it refers not only to "scaling" as a linear transformation that extends or reduces objects. We always consider the concept of "mathematics" which inseparably accompanies the scale. In subchapter 1.4, we discussed the relationship between physics and the underlying mathematics, emphasizing that mathematics is strongly deductive, while physical axioms are not strongly self-evident. Nevertheless, physicists derive new conclusions from those axioms, even with mathematical deduction. Seemingly, the term "scale" in physics includes a phenomenon's space or time characteristics and its description by a certain mathematical method; therefore, it is crucial to define scale in mathematics and physics properly. In other words, could any physical theory, abundantly described by a mathematical formalism, be formulated within the same "scale" where it would be possible to prove or refute it? By analogy with axioms, any question related to the theory has to be formulated within the same "scale." At this point, we come to the question of axiomatization in physics. Since this matter is not within the scope of this subchapter—that is, what we want to say about scaling, we consider it necessary to mention only the following: 1. The fact is that axiomatization in physics is very much work in progress. 2. In general, the axiomatic method is limited in physics because of "the requirement that all axioms used in a given theory of Physics subjected to axiomatization have a clear and explicit physical meaning, and thus be given by, and hence be within the realms of 'physical intuition'" [1].

We can now address the question of a scale in the axiomatized physical theory. If it is possible to create a formal system in a mathematical sense, then all theorems describe all phenomena in that theory. Correspondingly, we have the scale where we can prove or refute theorems. If a circle represents such an axiomatized physical theory symbolically, then the corresponding axiomatized mathematics may be applied in that circle. Within that circle, undecidable propositions may exist according to Gödel's incompleteness theorems. If we extend the axioms from the previous circle with a new circle, we reach a new scale or the circle encompassing the previous circle with its axioms and theorems. This is similar to the propagation of surface waves; therefore, we call this scaling *propagating scaling*.

Emergence in physics and transition between phases. In his blog, Daniel Little asks, "What is the connection between 'emergent phenomena' and systems that undertake phase transitions?" [2]. We discussed emergent properties in subchapter 1.3, covering a general story that applies to any complex system. Since we intend to include complex physical systems, the "specificity" of emergent properties in physics should be introduced somewhere in the text, which is done in this subchapter, which refers to scaling, universality, and phase transitions.

The weakly emergent properties of a complex system are properties that cannot be derived from the characteristics of components. In contrast, strongly emergent properties cannot, *in principle*, be derived from the full knowledge of the properties of components and states since they must be understood in their own terms. However, some questions and problems regarding the theory of emergence originate from physical facts and mathematics of *phase transitions*. Kivelson and Kivelson define the following: "An emergent behavior of a physical system is a qualitative property that can only occur in the limit that the number of microscopic constituents tends to infinity" [3]. The spontaneously broken symmetries which characterize distinct phases of matter are weakly emergent in the sense of this definition. Bedau [4] classifies some phase transitions as weak emergence.

A phase transition is a phenomenon in which the variation of external conditions (usually the temperature) induces an important change in a system: a completely new property, generally entitled *order parameter*, appears, and its existence defines a new phase, the ordered phase. The ordered phase usually exists at low temperatures, and a rise in temperature causes the order parameter to become smaller and vanish at a certain (transition) point. On the other hand, from the perspective of high temperatures, when the disordered phase that usually possesses a high symmetry reaches the transition point, the order parameter appears and lowers the symmetry, a phenomenon called spontaneous symmetry breaking. To be more precise, we now separate *first-order transitions* from *continuous transitions* belonging to the group of *critical phenomena* [5].

First-order transitions are characterized by the fact that the first derivative of the free energy of a system has discontinuity (the abrupt fall of the order parameter, the existence of latent heat). During continuous transitions, at the *transition point* or *critical point*, often characterized by the critical temperature T_c at which the order parameter changes smoothly like $|T - T_c|^{\beta}$, where β is the so-called *critical exponent*. Some other properties possess a singularity. For example, the specific heat for constant volume behaves like $|T - T_c|^{-\alpha}$, where the critical exponent is either α or α' depending on how we approach the critical point. To summarize, the mathematical form that relates physical quantities to its "distance" from the critical point is defined by power laws, introducing critical exponents. This is a concise introduction, and for further reading, one can consult enormous literature on this topic, but we prefer to quote probably the best book by Stanley [6], which dates back to times before the renormalization group appeared. Note that in critical phenomena, "scaling" has a particular meaning compared to the meaning we introduced at the beginning of this subchapter.

What is the relation between phase transitions and critical phenomena? Some authors treat them as synonyms, but the relationship is more complex. Some first-order transitions are not critical phenomena (water has both kinds of transitions). On the other hand, there are situations with critical behavior but not properly defined phases. For example, we may consider Anderson localization, percolation, topological transitions, and Mott transition to be critical phenomena but not phase transitions. This is why we need to define the difference between them.

Richard Solé provides a detailed description of the current state and elaborates his own ideas in the book *Phase Transitions* [7]. Like many complexity theorists, he observes phenomena with phase transitions in various systems, including *simple physical*, biological, and social systems. One of the remarks that can be made on the interpretation and modeling of phase transitions in many complex systems (such as physical, biological, and social) is the following: The fact that *unrest* (changes from the normal state) is not a new *equilibrium* phase of the substrate of detached individuals. It is rather an occasional anomalous state that exists briefly. Solé's idea developed in his book is that it is possible to understand phase transitions regardless of particular micro-level mechanisms. This assumption claims that certain complex systems have common dynamical characteristics, formal and abstract in some sense; therefore, they do not require the understanding of their underlying micro mechanisms. Consequently, it is sufficient for us to know that some system is formally similar, for example, to a two-dimensional array of magnetized atoms (the Ising model); then we can conclude that the phase-transition behavior of the system has specific mathematical properties. In summary, "system properties do not require derivation from micro dynamics" [2] is an assumption that does not shed new light on the topic of emergence *per se*. However, it is a matter of great importance for the practical research of phase transitions.

Scaling, scale invariance, and universality. What is the position of scaling in the physics of complex systems, especially when it comes to their properties, such as emergence and complexity? The common problem is to find an expression for the dependence between two measurable quantities (see subchapter 4.1). It is usually assumed that those two quantities depend on each other in a power law fashion. Experience tells us that not everything is covered by the power law in nature and the physical world, but the power law is an example of scaling laws that describe the functional relationship between two physical quantities scaling with each other over a considerable interval. One of the consequences of the properties of power laws is relations connecting critical exponents, which are often entitled "scaling laws" in literature.

Scaling laws have a property called *scale invariance*—that is, those laws do not change if different scales are multiplied by a common factor, which signifies *universality*. According to [5], "by universality is meant a tendency for the set of exponents found for diverse systems to partition themselves into distinct 'universality classes.'" In other words, all systems belonging to the same universality class have the same exponent. It can be said that both scaling invariance and universality are the pillars of the theory of phase transitions and critical phenomena [6, 8]. The understanding of these phenomena came after an extensive period of collecting empirical evidence and placing it in a phenomenological context using the principles of scale invariance and universality. Although theoretical developments exist, no unified, coherent theory satisfactorily describes empirical evidence at present. Used mathematical formalisms contribute significantly to this situation since they operate at different scales. The physical theory risks being flooded with numerous latent assumptions that can neither be proved nor refuted.

Universality in complex systems. If we move away from physical complex systems to study scale invariance in other complex systems, we can encounter numerous intriguing examples of universality. In [9], an illustration (Figure 2 in that paper) is a log-log plot showing the dependence of the standard deviation of a family of histograms for the growth rates of a group of bird species similar in size. The linearity of the curve implies a power law relation, and the slope of the regression fit gives the exponent β having the value $\approx 1/3$ that is roughly twice as large as that found for the firm growth and the country growth ($\beta \approx 1/6$) and slightly larger than that found for university research budgets ($\beta \approx 1/4$). It is seen that the fashion in which the standard deviation of a family of histograms for the growth rates of a group of bird species similar in size decreases with the size is qualitatively the same but quantitatively different from that for the economy examples (the power law). This "universality" is also found in scaling laws in financial studies, whose subject is a *par excellence* example of complex systems. There is a latent hope that, in the future, a theoretical framework will be established within which we can understand these intriguing examples of the universality of complex systems. Still, we should have in mind the question: Why does universality come up in these complex systems? Perhaps Solé's idea about the understanding of phase transitions regardless of microscopic mechanisms can be useful here. One should test how Solé's idea of the heuristic "explanation" of scale invariance and universality in the Ising model (which is simple) can be mapped by a specific mathematical language to other complex systems. Namely, the Ising model can be formulated in terms of the connectivity of geometrical objects (percolation). Note that this idea can be extended to biological and ecological systems by mapping their complex interactions onto some geometric system so that the scaling and universality features of complex systems may finally be understood—just as in the Ising model [9].

7.2 The Separation of Scales and Capabilities of the Renormalization Group

The separation of scales. In physics, the term "scale" indicates a range of possible lengths or time values. When a phenomenon is modeled across such a wide space-time range, a single model cannot capture it; therefore, this range is segmented into several scales, which is a core idea of the *separation of scales*. In its applications, this approach is a mathematical construct. One of the central issues in the physics of complex systems is the mathematical interpretation of interdependence among components. At present, it cannot be done with success because we rely on traditional mathematics with its limitations when interpreting these interactions (see subchapter 1.4). These limitations are related to the fact that models only apply to systems where it is possible to separate behavior between micro and macro scales. Otherwise, interactions among entities that cause behavior across scales violate this separation. Therefore, to make a larger step forward, new mathematics for complex systems is needed to capture their internal interactions better. Still, efforts in modeling complex systems are only made to obtain information at a larger scale. Since the scale of description and scale of interactions are similar, we can consider the large-scale impacts of the environment on a system (and reciprocally) if we have the description of large-scale behavior [10].

An example of the separation of scales we deal with in this subchapter is the *renormalization group* (RG). The first step in the RG procedure is the separation of scales. Afterward, the degrees of freedom of "lower" scales are eliminated so that the principal dynamic characteristic of a system, the Hamiltonian, maintains the same form but with different parameters. By iterating this procedure, we follow the change of parameters until a value that no longer changes is reached, which is the point at which physics intervenes. From a mathematical standpoint, we reached the *fixed point* that enables us to determine the critical exponent, and we will discuss it later.

The core history of RG. When the renormalization group "celebrated" its fiftieth birthday in 2001, Dmitry Shirkov published an article concisely describing its short history [11]. Quantum field theory (QFT) made an advance with RG in late 1940; from then on, new theoretical developments were mostly based on renormalization procedures. "The Normalization Group in Quantum Theory" [12] was the first paper published on this topic but remained unnoticed. In the mid-1950s, the renormalization group improved approximate solutions to QFT equations and the study of singular behavior in ultraviolet and infrared limits. This method was later transferred from QFT to quantum statistics for the analysis of phase transitions and further to other fields of theoretical and mathematical physics. A clearer formulation was needed, in fact in QFT, finite renormalization transformations form a continuous group (the Lie group) for which Lie's differential equations hold [13]. A more transparent picture was given in the papers by the Russian physicists Nicolai Bogoliubov and Dmitry Shirkov (1955–1956). In two short notes [14, 15], they established the connection between Stückelberg and Petermann's work and Gell-Mann and Low's work [16]. These results were published in a monograph [17]. Afterward, RG methods became an essential tool in the quantum field analysis of asymptotic behavior. Note that the term "renormalization group" and the central notion of the RG method algorithm were first introduced in the papers mentioned above by Bogoliubov and Shirkov. Perhaps the most important physical result obtained by the RG method was the discovery of the "asymptotic freedom" of non-Abelian vector models, which was elaborated in [18, 19]. We finalize this core history of RG with Kenneth Wilson's contribution [20], who developed a specific version of the RG formalism for statistical systems. Namely, based on the idea of "blocking" (averaging over a small part of a big system), he obtained a discrete semigroup of blocking operations. Wilson's renormalization group was then used for calculating critical indices in phase transitions and was applied to polymers, percolation, noncoherent dynamical chaos, radiation transfer, and other problems. For these works, Kenneth Wilson was awarded the Nobel Prize in 1982.

A short guide through RG. Renormalization techniques are applied in the different domains of theoretical physics and, recently, applied physics, medicine, economy, etc., and the use of the same term "renormalization" might be confusing. Physicists study how a theory transforms under a family of operations that typically constitute a mathematical group. Those operations are either discrete or continuous. In the case of RG, physicists consider how the theory transforms under an operation that "scales" a unit of length. Considering such scaling as a type of lossy data compression or smoothing seems suitable. Consequently, RG theory can be viewed as a metatheory that describes how theories change under lossy data compression [21]. From that point of view, let $b_1, b_2, b_3 \geq 1$. If R_b denotes data compression with the compression factor b of a theory, then $R_{b_1 b_2} = R_{b_1} R_{b_2}$. Note that $R_1 R_b = R_b$ (identity property) and $R_{b_1}(R_{b_2} R_{b_3}) = (R_{b_1} R_{b_2}) R_{b_3}$ (associative property). However, we cannot set $R_{1/b} R_b = R_1$ (inverse property) because $1/b$ is not greater than 1. In a physical sense, the reason why RG operations do not have inverses is that they represent lossy data compression that is irreversible. Therefore, the set of the transformation $\{R_b : b \geq 1\}$ is not really a mathematical group. It only satisfies the properties of a weaker algebraic structure called a semigroup that might or might not lead to one or more fixed points.

Many books and papers on RG theory are available, so we advise readers to examine other sources, especially when considering a specific RG-related problem. Apart from the core papers mentioned in this subchapter, we recommend the books that introduce this subject to the reader gradually: (1) the book for beginners written intuitively by William McComb [22], (2) the book (slightly advanced than McComb's book) written by Nigel Goldenfeld [5], (3) the book about quantum field theory from the perspective of high energy physics by Mark Srednicki [23]. We also recommend the following papers: (1) an educationally written paper by Shang-Keng Ma [24], (2) the paper by Annick Lesne [25] that is a smooth connection of RG and mathematics, and (3) the paper by Margaret Morrison [26] in which RG is considered from the standpoint of the philosophy of science.

The applications of RG theory can be found in various scientific areas. Undoubtedly, it will be extended to areas that will arise in the future. Here we list just some of the current domains: fluid mechanics (turbulence), chaos and fractal theory, differential equations, statistical mechanics (phase transitions), probability and statistics, quantum field theory, both relativistic (high energy) physics and nonrelativistic (condensed matter) physics, classical Shannon information theory, quantum information science, quantum Shannon information theory, the study of quantum entanglement, medicine (the RG approach to pandemics—case study COVID-19), economy (the RG scaling of stock markets), etc.

Scaling behavior may often be described in the context of theoretical ideas roughly grouped under the term *renormalization*. It describes how parameters specifying a system must be adjusted under the assumed changes of basic dynamics. This should be done to not modify measurable properties at length or time scales of interest. In other words, physicists try to understand how different length scales in physics correlate with each other or how to endlessly connect the micro world with the macro world. In these renormalization flows, a simple hypothesis of a fixed point's existence is sufficient to describe the appearance of universal scaling laws qualitatively. Nevertheless, in nature, most complex systems cannot be captured by a renormalization approach. RG theory seems to be much more "meta" than most theories in physics. It describes how theories change under lossy data compression [21].

We should distinguish *renormalization* applied for *regularization*, mostly in QFT to remove UV divergences, from a *renormalization flow used* to explore critical phenomena and universal scaling properties. Note that the relation between QFT and critical phenomena and their corresponding renormalizations are discussed in [27] and elaborated in Jean Zinn-Justin's book comprehensively [28]. There exists a mathematically precise, usually accepted definition of the term "regularization procedure." In contrast, no precise and generally accepted definition of this term exists in perturbative QFT—just various regularization schemes with their advantages and disadvantages. In other words, regularization introduces some scale into the theory so that certain amplitudes that were divergent before become finite afterward.

Feigenbaum's RG equation for the period-doubling bifurcation sequence. The application of RG theory has solved many problems since its appearance, and a lot has been written about it. We consider an application of RG to dynamical systems by using a simple one-dimensional dynamical system represented by the logistic equation (3.3.2) that will be written in the form $X_{i+1} = f_r(X_i)$. The subscript denotes the dependence of f on the logistic parameter r. For a specified initial value X_0 of the variable X, the logistic equation generates a sequence of values $\{X_1, X_2, X_3, \ldots\}$. The index represents discrete time. We consider the properties of these sequences over a long period of time shortly—that is, the behavior of the logistic equation changes as i varies. One exciting property of nonlinear dynamical systems and the logistic equation is that they can exhibit universality. This feature of dynamical systems was first recognized by Feigenbaum [29] in the period-doubling route to chaos. In the form of a very short summary, we present a derivation of Feigenbaum's RG equation relaying on the main point in [30] used to understand this universality. Feigenbaum observed that the range r in which an orbit of length 2^n is observed (Δr) diminishes geometrically with n when n is large: $\delta = \Delta r_{n-1} / \Delta r_n$ = 4.6692... Moreover, he explored several different maps finding (for all of them) that the range of the logistic parameter as the function of n obeys the logistic equation with the same value of δ. He also exposed that the values of X on orbits have universal properties. The orbits of length 2^n in the period-doubling sequence include the value $X = 1/2$ when the function $f_r(X)$ reaches its maximum value. He found that for a given cycle of length 2^n

$$\frac{\frac{1}{2} - f_{r_n}^{2^{n-1}}\left(X = \frac{1}{2}\right)}{\frac{1}{2} - f_{r_{n+1}}^{2^n}\left(X = \frac{1}{2}\right)} \approx -\alpha, \tag{7.2.1}$$

where $\alpha = 2.502\,907\,875\ldots$ It is seen that many different map functions display this scaling with exactly the same value of α.

Feigenbaum [31] demonstrates that the exponents α and δ are universal over a large class of systems [30] and are related to the existence of a universal function $g(z)$ satisfying

$$g(z) = -\alpha g\left(g\left(\frac{z}{\alpha}\right)\right). \tag{7.2.2}$$

The universality of exponents comes from the following fact: when $g(z)$ has the quadratic maximum $[g(z) = a_0 - a_0 z^2 + \ldots]$, the Equation (7.2.2) determines α and δ uniquely. The Equation (7.2) was derived by defining an infinite sequence of functions, where $g(X)$ was defined as the limit of this sequence. The reader can find methods for solving the RG Equation (7.2.2) and obtaining $g(z)$, α, and δ in numerous references. The essence of the derivation in [30] is identical to that used to derive the RG equations for second-order phase transitions. In fact, his simpler method for the period-doubling in 1D maps is analogous to the decimation RG for the one-dimensional Ising model.

How can we summarize the history of RG and its contributions to physics but also to other sciences? It is considered to be a profound epistemological shift. Since physics is more or less based on successful heuristic models, the contribution of RG theory to physics should be considered in that context: 1. It has done a lot of work since its appearance. 2. It helped describe many physical phenomena in a way that initially seemed unusual. However, the fortunate circumstance was that physicists had a clear insight into a macroscopic level of the phenomenon obtained by an experiment. Then they followed a strong theoretical procedure to describe that phenomenon going toward the microscopic world. 3. Currently, RG theory over correlation functions scale and correlation functions is crucial in the study of quantum entanglement and both classical and quantum Shannon information theory. 4. It seems that RG theory is a sophisticated theory in which mathematics is rather dominant, while its keystone is dimensional analysis.

Subchapter 7.4, dedicated to complexity and time scale, offers examples of the separation of time scales. Let us mention another example, so-called self-organized criticality. A typical example is sand pile automata, where one time scale dictates how sand grains drop on the pile. At the same time, this process causes instability and avalanches occurring at a completely different time scale. An extensive review of such systems can be found in [32] and references therein.

7.3 A Phase Transition Model Example: The Longevity of the Heisenberg Model

We explained elementary terminology about a phase transition in the previous subchapter. The order parameter for magnetic systems is spontaneous magnetization (resulting average magnetic moment in the absence of a magnetic field) in the ferromagnetic (ordered) phase. With a temperature rise, magnetization vanishes at the so-called Curie point, which is actually the critical temperature. We already recommended the book by Stanley [6]. The approach to the theory of phase transitions is to start from the free energy of a system and derive from it all important properties looking for singular behavior—for example, in susceptibility.

Two possible approaches to modeling are microscopic and macroscopic (see subchapters 4.1 and 4.2). One can either formulate a macroscopic, phenomenological model of free energy based on symmetries or start from a microscopic model and evaluate the statistical sum and free energy from it. We concentrate on an example of the second approach. Two mathematical models are the pillars of the application of statistical mechanics in the magnetism of localized magnetic moments: the Ising model [33] and the Heisenberg model [34]. They are both designed to describe the phase transition into the ferromagnetic phase. These models describe the interaction of magnetic moments, more precisely spins. The Ising model considers the interaction of z components of the spin 1/2. Therefore, it is often considered one of the simplest statistical models showing the phase transition. Ernst Ising found that there was no phase transition for the one-dimensional case in 1925 [33], while in 1944, Lars Onsager gave an elaborated analytical solution for the more complicated two-dimensional square lattice Ising model that includes the phase transition [35]. The analytical solution for the three-dimensional Ising model does not exist yet. In the Heisenberg model, the spins of magnetic systems are treated quantum mechanically. A key difference between the Ising model (IM) and the Heisenberg model (HM) is that IM has a discrete symmetry (Z symmetry), while HM has a continuous (rotational) symmetry.

The Heisenberg model is based on the expression for the contribution to the energy of the system arising from a particular property of quantum mechanical systems: symmetry due to the indistinguishability of identical particles. This contribution can be expressed in terms of particle spins, and Heisenberg generalized it to represent the energy of the interaction between the system of localized ions having magnetic moments.

The expression for the energy of a system is

$$H = -1/2I \sum \vec{S}_m \circ \vec{S}_n. \tag{7.3.1}$$

Here $I > 0$ is the so-called exchange interaction having the dimension of energy, while the spin operators $\vec{S}_m$ and $\vec{S}_n$ are dimensionless to facilitate calculations. The subscript denotes the lattice site, and, in the simplest case, only the nearest neighbors are considered. This form corresponds to a ferromagnet since the ground state corresponding to the lowest energy is obtained when all spins are parallel (without the minus sign, the system describes an antiferromagnet). This is the case of equal spins, but if they are different, we deal with the ferrimagnet. Transition is described in the following manner: In the ground state, all spins are aligned. With a temperature rise, spins begin to deviate from their directions; therefore, we have local magnetization fluctuations. Areas with fluctuations become larger (long-range fluctuations), and at the critical temperature, no order exists.

One of the first steps is to look for some exact statements, if any. Here we have in mind the theorem [36] that states that in an isotropic model, an ordering at the temperature $T > 0$ in dimensions 1 and 2 cannot exist since fluctuations are so strong that they prevent ordering. Therefore, the only way to describe such situations existing in practical applications (i.e., to achieve the magnetization in (quasi-)low dimensional systems) is to assume some disturbance of ideality, such as additional coupling or the anisotropy of the Hamiltonian.

Calculation techniques are various [37]. The first step, usually preferred by experimental physicists, in the interpretation of results is to introduce the so-called effective ("mean") field [38], where the Hamiltonian is substituted with the Hamiltonian of the system of noninteracting spins in an effective magnetic field that is proportional to magnetization. The further step is to write down the equations of motion for spin operators where the chain of equations is truncated at some step, and the degree of approximation depends on it. This leads to self-consistent equations for magnetization. Realistic systems cannot usually be described by a model completely, so many items are added, which opens a broad field to adapt HM for the explanation of situations occurring in practical applications [39–43].

Another important topic is behavior at low temperatures. When energy is added to the system that is in the ground state, spins start to deviate from their perfect ordering in a privileged (magnetization) direction, and so magnetization deviates from its maximal value. This excitation is transmitted through a crystal due to translational symmetry, and we observe so-called spin waves. Dyson's brilliant work opened the door to many generalizations [44–45]. At low temperatures, spin operators can be substituted with so-called Bose operators whose statistics is well elaborated ([46–50] among many others).

Since the 1980s, interest in so-called solitary waves or solitons has arisen. They have the fascinating property of colliding without interaction, which makes them suitable candidates for energy transport without losses. Theoretical and even further experimental studies indicated that such excitations could exist in systems described by HM. The study demanded the transition to classical quantities, and much work has been done by using so-called coherent states to perform the transfer [51–53].

Why did we decide to present the Heisenberg model in this subchapter and enhance its longevity? Because it simply acts like a "black box" to which a lot of material can be added to adapt it to our needs. Therefore, it is possible to obtain many useful and also practical results because of the great experimental possibilities and the power of computational physics. One generalization, the so-called "n model" [6], allowed an unexpected breakthrough in physics.

Pierre-Gilles de Gennes demonstrated that in the peculiar limit of vanishing n, critical properties could be mapped onto those of polymer solutions [54]. More recently, the Heisenberg model became a valuable tool for the study of a particular sort of quantum computer (see the lectures and papers from the 4th School and Conference on Spin-based Quantum Information Processing [55]).

Finally, the reader will note that the text in this subchapter is suitable for nonexperts in the field. We think that today in science, it is necessary for scientists to be "quickly" informed about many topics, and this text is an example of such information.

7.4 Complexity and Time Scale: The "Breaking" Point with an Experimental Example

A time scale and "breaking" point in complex systems. Many complex systems undergo sudden transitions from one dynamical state to another. Complexity in such systems comes from nonlinearity in the dynamics of interacting subsystems, their heterogeneity, and the presence of multiple *time scales* in their dynamics (see subchapter 3.4). An example of such a system is described in subchapter 4.5—the coupled patches of vegetation and connected desert in grid-cell (Figure 4.3) consisting of a subsystem among many others in climate models [56]. Most complex systems are intrinsically dynamical in nature, and their time evolution is characterized by two points: the *threshold* and the *tipping point*. Some change should be expected when a complex system passes any of these points. It is known that in such systems, thresholds are the most important sources of nonlinear behavior. Although the tipping point has a similar meaning to the threshold, it is more related to systems as a whole. After passing that point, the system shifts fundamentally into a different equilibrium state. Detecting and predicting these points is the most important task in complex systems theory. One way to detect the point at which the state of a complex system changes is to calculate the running mean of Kolmogorov complexity (see subchapter 6.1) as an indicator of the state of a complex system. By pursuing the system's complexity from the time series obtained by measurement and increasing the window size for which the running Kolmogorov complexity (RKC) is calculated, we obtain the average complexity values for the whole period. If there exists a point at which a significant drop in complexity begins, then that point is called the "breaking" point (Figure 7.1), which is discussed in great detail in this subchapter. It signifies that the system's state is changed for some reason—for example, urban activities in river streamflow, errors in the construction of detectors for gamma radiation and neutrons, changes caused by brain disorders, etc.

The Separation of Scales in Complex Systems 109

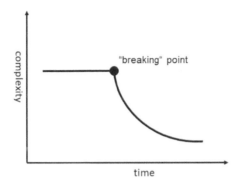

FIGURE 7.1
Toward the "breaking" point that can occur in the time evolution of complex systems.

Complexity and "breaking" point. The method known as RG contributed to the descriptions of phenomena in physics; however, it is neither exact nor completely controlled. Instead, it should be regarded as conceptual in principle. The "power" of RG is its application relying fundamentally only on scaling that can immediately be tailored for a particular application. Regarding an experiment, we can pose the question of how RG theory's predictions match experiments.

In the previous subchapters, we noted that RG could be considered through scaling as a type of lossy data compression that causes data loss using inexact approximations. However, any "intervention" similar to this mostly damps complexity—one of the most distinguishing properties of complex systems, and one that cannot be modeled. We can only obtain information from a physical phenomenon by making measurements or by a time series that contains the time evolution of complexity calculated by one of the complexity measures. Why is the time evolution of information important? Because it can show the point after which a smaller or larger drop of complexity in a system exists, or the "breaking" point (Figure 7.1). This point indicates a close interaction between the surrounding environment and the complex system.

Perhaps the best way to illustrate the "breaking" point is to visualize the "overall" dynamics of the logistic equation when its parameter ranges from 3.6 to 4. That dynamics can be represented by the Lyapunov exponent and the running Kolmogorov complexity (the upper panel of Figure 7.2). Almost the entire domain of the system is characterized by high chaos and mostly low complexity; however, the regions of high complexity are also visible, which indicates the presence of the "breaking" point. The lower panel of Figure 7.2 shows the position of "breaking" points even more clearly—around 90 percent of RKC values lie on an almost straight line parallel to the x-axis with values around 0.15. All other points have RKC values from about 0.35 to 0.80, which specifies "breaking" points.

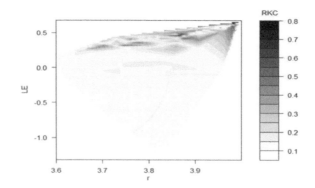

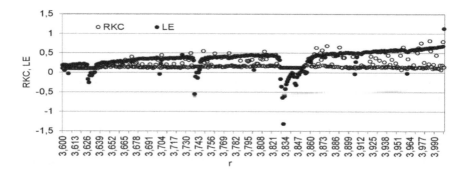

FIGURE 7.2
2D map of the running Kolmogorov complexity (RKC) and Lyapunov exponent (LE) (*upper*) and LE and RKC graphs (*lower*) of time series of the logistic equation versus the logistic parameter *r*. Calculations were performed for 3000 time series, initial condition $x_0 = 0.35$, and fixed windows with different sizes (200, 300, 400, ..., 2800).

The impression is that this procedure, the application of the running Kolmogorov complexity and extension of the size of a time window, is analogous to that used in RG theory. However, it is now done by extracting points at the time scale, in the time evolution of complex systems, where complexity usually starts to decrease. Note that the lower panel of Figure 7.2 does not show the time determination of the "breaking" point but only its value for the running Kolmogorov complexity for a certain value of the logistic parameter.

An experimental example. Let us see how the "breaking" point appears in the time series of neutron and gamma rays measured with organic scintillators [57]. Figure 7.3 shows the time distributions of neutron and gamma-rays' events for the plutonium metal plates (8 and 16 stilbene detectors) [58]. It is seen that the running complexity of PAHN_7 is nearly constant, while the "breaking" point occurred in the running complexity of PAHN_3. Note that

The Separation of Scales in Complex Systems

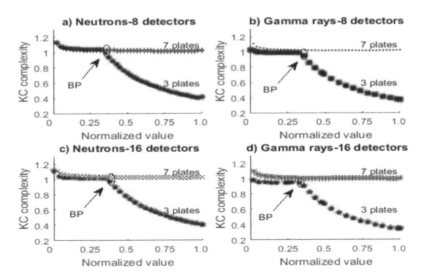

FIGURE 7.3
The running Kolmogorov complexity of the neutron and gamma time series with 3 and 7 PAHN plates. The "breaking" points are rounded by ellipses. The position of the sample on the x-axis is normalized by its time series length. BP is the notation for the "breaking" point. Values on the time axis are normalized by the highest value in the time series. (Reproduced by permission from [57].)

for the time series with the higher number of Pu plates, the complexities were also high but were not drawn. The occurrence of the "breaking" point is a typical situation related to the interaction between radiation and the environmental interface. If that type of interaction exists, the "breaking" point is not noticeable for smaller size of time windows, but for larger sizes, complexity starts to decrease rapidly at some point that is exactly the "breaking" point. The sharp drop of complexity is not related to the nuclear process itself but rather to the neutron and gamma-rays' interaction with the detector. The source of the unproper detection can be attributed to the architecture of the detector's configuration and the influence of the surrounding environment.

References

[1] E. E. Rosinger, "A disconnect: Limitations of the axiomatic method in physics," unpublished. Available: https://www.researchgate.net/publication/281159338_A_Disconnect_Limitations_of_the_Axiomatic_Method_in_Physics

[2] D. Little. "Understanding society." *Understandingsociety.blogspot.com.* https://understandingsociety.blogspot.com/2016/04/phase-transitions-and-emergence.html (accessed Jul. 5, 2022).

[3] S. Kivelson and S. A. Kivelson, "Defining emergence in physics," *npj Quantum Mater.*, vol. 1, Nov. 2016, Art. no. 16024, doi:10.1038/npjquantmats.2016.24.

[4] M. A. Bedau, "Weak emergence," *Noûs*, vol. 31, pp. 375–399, 1997.

[5] N. Goldenfeld, *Lectures on Phase Transitions and the Renormalization Group*. Reading, MA, USA: Addison-Wesley, 1992.

[6] H. E. Stanley, *Introduction to Phase Transitions and Critical Phenomena*. New York, NY, USA: Oxford University Press, 1971.

[7] R. V. Solé, *Phase Transitions*. Princeton, NJ, USA: Princeton University Press, 2011.

[8] H. E. Stanley, "Scaling, universality, and renormalization: Three pillars of modern critical phenomena," *Rev. Mod. Phys.*, vol. 71, no. 2, pp. S358–S366, Mar. 1999, doi:10.1103/RevModPhys.71.S358.

[9] H. E. Stanley, L. A. N. Amaral, P. Gopikrishnan, P. C. Ivanov, T. H. Keitt, and V. Plerou, "Scale invariance and universality: Organizing principles in complex systems," *Physica A*, vol. 281, no. 1–4, pp. 60–68, Jun. 2000, doi:10.1016/S0378-4371(00)00195-3.

[10] Yaneer Bar-Yam. "Separation of scales: Why complex systems need a new mathematics?." *Necsi.edu.* https://necsi.edu/11-separation-of-scales/ (accessed Jul. 10, 2022).

[11] D. V. Shirkov. "Fifty years of the renormalization group." *Cerncourier.com.* https://cerncourier.com/a/fifty-years-of-the-renormalization-group/ (accessed Jul. 10, 2022).

[12] E. C. G. Stückelberg and A. Petermann, "The normalization group in quantum theory," *Helv. Phys. Acta*, vol. 24, no. 4, pp. 317–319, 1951. (proveriti jezik).

[13] E. C. G. Stückelberg and A. Petermann, "Normalization of constants in quantum theory," (in French), *Helv. Phys. Acta*, vol. 26, no. 5, pp. 499–520, 1953.

[14] N. N. Bogoliubov and D.V. Shirkov, "On the renormalization group in QED," (in Russian), *Dokl. Akad. Nauk*, vol. 103, pp. 203–206, 1955.

[15] N. N. Bogoliubov and D.V. Shirkov, "Lee type model in QED," (in Russian), *Dokl. Akad. Nauk*, vol. 105, pp. 685–688, 1955.

[16] M. Gell-Mann and F. E. Low, "Quantum electrodynamics at small distances," *Phys. Rev.*, vol. 95, no. 5, pp. 1300–1312, Sep. 1954, doi:10.1103/PhysRev.95.1300.

[17] N. N. Bogoliubov and D. V. Shirkov, *Introduction to the Theory of Quantized Fields* (Monographs in Physics and Astronomy 3), G. M. Volkoff, Trans., New York, NY, USA: Interscience, 1959.

[18] D. J. Gross and F. Wilczek, "Ultraviolet behavior of non-Abelian gauge theories," *Phys. Rev. Lett.*, vol. 30, no. 26, pp. 1343–1346, Jun. 1973, doi:10.1103/PhysRevLett.30.1343.

[19] H. D. Politzer, "Reliable perturbative results for strong interactions?," *Phys. Rev. Lett.*, vol. 30, no. 26, pp. 1346–1349, Jun. 1973, doi:10.1103/PhysRevLett.30.1346.

[20] K. G. Wilson, "Renormalization group and critical phenomena. I. Renormalization group and the Kadanoff scaling picture," *Phys. Rev. B*, vol. 4, no. 9, pp. 3174–3183, Nov. 1971, doi:10.1103/PhysRevB.4.3174.

[21] R. R. Tucci. "Honey, I shrunk the theory." *Ar-tiste.com.* http://www.ar-tiste.com/honey-i-shrunk-the-theory.pdf (accessed Jul. 13, 2022).

[22] W. D. McComb, *Renormalization Methods: A Guide for Beginners*. New York, NY, USA: Oxford University Press, 2007.
[23] M. Srednicki, *Quantum Field Theory*. Cambridge, UK: Cambridge University Press, 2007.
[24] M. Shang-Keng, "Introduction to the renormalization group," *Rev. Mod. Phys.*, vol. 45, no. 4, pp. 589–614, Oct. 1973, doi:10.1103/RevModPhys.45.589.
[25] A. Lesne, "Regularization, renormalization, and renormalization groups," in *Vision of Oneness*, I. Licata and A. J. Sakaji, Eds., Rome, Italy: Aracneeditrice, 2011, pp. 121–154.
[26] M. Morrison, "Complex systems and renormalization group explanations," *Philos. Sci.*, vol. 81, no. 5, pp. 1144–1156, Dec. 2014, doi:10.1086/677904.
[27] K. G. Wilson, "The renormalization group: Critical phenomena and the Kondo problem," *Rev. Mod. Phys.*, vol. 47, no. 4, pp. 773–840, Oct. 1975, doi:10.1103/RevModPhys.47.773.
[28] J. Zinn-Justin, *Quantum Field Theory and Critical Phenomena (International Series of Monographs on Physics 113)*, 4th, ed. Oxford, UK: Clarendon, 2002.
[29] M. J. Feigenbaum, "Quantitative universality for a class of nonlinear transformations," *J. Stat. Phys.*, vol. 19, no. 1, pp. 25–52, Jul. 1978, doi:10.1007/BF01020332.
[30] S. N. Coppersmith, "A simpler derivation of Feigenbaum's renormalization group equation for the period-doubling bifurcation sequence," *Amer. J. Phys.*, vol. 67, no. 1, pp. 52–54, Jan. 1999, doi:10.1119/1.19190.
[31] M. J. Feigenbaum, "The universal metric properties of nonlinear transformations," *J. Stat. Phys.*, vol. 21, no. 6, pp. 669–706, Dec. 1979, doi:10.1007/BF01107909.
[32] B. Tadić and R. Melnik, "Self – organized critical dynamics as a key to fundamental features of complexity in physical, biological and social networks," *Dynamics*, vol. 1, no. 2, pp 181–197, Oct. 2021, doi:10.3390/dynamics1020011.
[33] E. Ising, "Contribution to the theory of ferromagnetism," (in German), *Z. Phys.*, vol. 31, no. 1, pp. 253–258, Feb. 1925, doi:10.1007/BF02980577.
[34] W. Heisenberg, "On the theory of ferromagnetism," (in German), *Z. Phys.*, vol. 49, no. 9–10, pp. 619–636, Sep. 1928, doi:10.1007/BF01328601.
[35] L. Onsager, "Crystal statistics. I. A two-dimensional model with an order-disorder transition," *Phys. Rev.*, vol. 65, no. 3–4, pp. 117–149, Feb. 1944, doi:10.1103/PhysRev.65.117.
[36] N. D. Mermin and H. Wagner, "Absence of ferromagnetism or antiferromagnetism in one- or two-dimensional isotropic Heisenberg models," *Phys. Rev. Lett.*, vol. 17, no. 22, pp. 1133–1136, Nov. 1966, doi:10.1103/PhysRevLett.17.1133.
[37] S. V. Tyablikov, *Methods in the Quantum Theory of Magnetism*. New York; Berlin, Germany; Vienna, Austria: Springer-Verlag, 1967.
[38] J. S. Smart, *Effective Field Theories of Magnetism*. Philadelphia, PA, USA: Saunders, 1966.
[39] M. Pavkov, S. Stojanović, D. Kapor, M. Škrinjar, and D. Nikolić, "Influence of the biquadratic interaction to magnetic surface reconstruction", *Phys. Rev. B*, vol. 60, no. 9, pp. 6574–6583, Sep. 1999, doi:10.1103/PhysRevB.60.6574.
[40] M. Pavkov, M. Škrinjar, D. Kapor, M. Pantić, and S. Stojanović, "Magnetic properties of antiferromagnetic bilayers analyzed in the spin and boson pictures, " *Phys. Rev. B*, vol. 65, no. 10, Feb. 2002, Art. no. 104512, doi:10.1103/PhysRevB.65.104512.

[41] M. Pavkov, M. Škrinjar, D. Kapor, M. Pantić, and S. Stojanović, "Influence of spatial anisotropy on the magnetic properties of Heisenberg magnets with complex structures," *Phys. Rev. B*, vol. 65, no. 13, Mar. 2002, Art. no. 132411, doi:10.1103/PhysRevB.65.132411.

[42] D. Kapor, M. Pavkov, M. Manojlović, M. Pantić, M. Škrinjar, and S. Stojanović, "Some peculiar properties of the complex magnetic systems," *Phys. Status Solidi B Basic Res.*, vol. 241, no. 2, pp. 401–410, Feb. 2004, doi:10.1002/pssb.200301929.

[43] D. Kapor, M. Pantić, M. Manojlović, M. Škrinjar, S. Stojanović, and M. Pavkov, "Thermodynamic properties of magnetic bilayers," *J. Phys. Chem. Solids*, vol. 67, no. 4, pp. 698–704, Apr. 2006, doi:10.1016/j.jpcs.2005.10.176.

[44] F. J. Dyson, "General theory of spin-wave interactions," *Phys. Rev.*, vol. 102, no. 5, pp. 1217–1230, Jun. 1956, doi:10.1103/PhysRev.102.1217.

[45] F. J. Dyson, "Thermodynamic behavior of an ideal ferromagnet", *Phys. Rev.*, vol. 102, no. 5, pp. 1230–1244, Jun. 1956, doi:10.1103/PhysRev.102.1230.

[46] D. Kapor, M. Škrinjar, and S. Stojanović, "Magnetic excitations in the semi-infinite Heisenberg ferromagnet with biquadratic exchange," *Phys. Lett. A*, vol. 192, no. 5–6, pp. 413–420, Sep. 1994, doi:10.1016/0375-9601(94)90229-1.

[47] M. Pavkov, M. Škrinjar, D. Kapor, and S. Stojanović, "Elementary excitations and low-temperature behavior of the Heisenberg magnet with three or four sublattices," *Phys. Rev. B*, vol. 62, no. 10, pp. 6385–6392, Sep. 2000, doi:10.1103/PhysRevB.62.6385.

[48] M. Pantić et al., "Low-temperature properties of ferromagnetic Fibonacci superlattices," *Eur. Phys. J. B*, vol. 59, no. 3, pp. 367–373, Oct. 2007, doi:10.1140/epjb/e2007-00298-8.

[49] M. Manojlović, M. Pavkov, M. Škrinjar, M. Pantić, D. Kapor, and S. Stojanović, "Spin-wave dispersion and transition temperature in the cuprate antiferromagnet $La_2 CuO_4$," *Phys. Rev. B*, vol. 68, no. 1, Jul. 2003, Art. no. 014435, doi:10.1103/PhysRevB.68.014435.

[50] S. M. Radošević, M. R. Pantić, M. V. Pavkov-Hrvojević, D. V. Kapor, "Magnon energy renormalization and low-temperature thermodynamics of O(3) Heisenberg ferromagnets, " *Ann. Phys.*, vol. 339, pp. 382–411, Dec. 2013, doi:10.1016/j.aop.2013.09.012.

[51] D. V. Kapor, S. D. Stojanović, and M. J. Škrinjar, "Semi-classical quantisation of an anisotropic exchange spin-field model," *J. Phys. C Solid State Phys.*, vol. 19, no. 16, pp. 2963–2968, Jun. 1986, doi:10.1088/0022-3719/19/16/016.

[52] M. J. Škrinjar, D. V. Kapor, and S. D. Stojanović, "The classical limit for the Holstein-Primakoff representation in the soliton theory of Heisenberg chain," *J. Phys. Condens. Matter*, vol 1, no. 4, pp. 725–732, Jan. 1989, doi:10.1088/0953-8984/1/4/007

[53] D. V. Kapor, M. J. Škrinjar, and S. D. Stojanović, "Relation between spin-coherent states and boson-coherent states in the theory of magnetism," *Phys. Rev. B*, vol. 44, no. 5, pp. 2227–2230, Aug. 1991, doi:10.1103/PhysRevB.44.2227.

[54] P.-G. de Gennes, *Scaling Concepts in Polymer Physics*. Ithaca, NY, USA: Cornell Univ. Press, 1979.

[55] "5 – Spin Qubit 4 – Konstanz – Sep 2018 – Spin-NANO." *Sheffield.ac.uk*. http://spin-nano.sites.sheffield.ac.uk/network-events/5---spin-qubit-4---konstanz (accessed Jul. 20, 2022).

[56] D. T. Mihailović, I. Balaž, and D. Kapor, *Time and Methods in Environmental Interfaces Modelling: Personal Insights* (Developments in Environmental Modelling 29). Amsterdam, The Netherlands; New York, NY, USA: Elsevier, 2016.

[57] D. T. Mihailović, S. Avdić, and A. Mihailović, "Complexity measures and occurrence of the "breakpoint" in the neutron and gamma-rays time series measured with organic scintillators," *Radiat. Phys. Chem.*, vol. 184, Jul. 2021, Art. no. 109482, doi:10.1016/j.radphyschem.2021.109482.

[58] A. Di Fulvio et al., "Passive assay of plutonium metal plates using a fast-neutron multiplicity counter," *Nucl. Instrum. Methods Phys. Res.*, vol. 855, pp. 92–101, May 2017, doi:10.1016/j.nima.2017.02.082.

8

The Representation of the Randomness and Complexity of Turbulent Flows

8.1 The Randomness of Turbulence in Fluids

We could not get rid of the impression that this chapter should be a synthesis of what Andrey Kolmogorov (1903–1987) did in areas belonging to the foundations of mathematics and hydrodynamics between 1933 and 1962. He was a Russian mathematician who made an impact on probability theory, turbulence, and many other mathematical disciplines. Like Kurt Gödel, his was a mathematical mind who saw what many mathematicians and physicists failed to see.

Turbulence. Turbulence is still one of the unsolved problems in classical physics, since the equations of the motion of turbulent flows, impeccably known as Navier-Stokes equations and primarily driven by their nonlinearity, eliminate any hope of finding analytical solutions. A distinction criterion between turbulent and laminar flows is a crucial issue in turbulence theory. Interestingly, Osborne Reynolds (1842–1912), the pioneer in the research of turbulence, did not make use of the terms "laminar" or "turbulent." Most textbooks characterize laminar flow as the condition in which fluid flows in an orderly fashion whereas turbulent flow is the condition in which fluid flows are irregularly mixed. It appears that it was Julius Oscar Hinze (1907–1993) who first gave a detailed definition of turbulence: "Turbulent fluid motion is an irregular condition of flow in which the various quantities show a random variation with time and space coordinate, so that statistically distinct average values can be discerned" [1]. Afterward, many authors did not describe turbulence similarly; rather, they simply defined it indirectly by enumerating its nature [2]. What all those attempts to define turbulence had in common is their observation that turbulence was an irregular and random phenomenon, but no one explained just what was meant by the term random. The Reynolds number is defined as $Re = ul/\nu$, where u is the flow velocity, l is the characteristic linear dimension, and ν is the kinematic viscosity of the fluid (the laminar regime Re is up to 2300; the transition regime 2300

Re < 4000; the turbulent regime Re > 4000; these values are for a pipe). Here we point out that the depth of its definition is related to determining whether the type of the flow pattern is laminar or turbulent.

Kolmogorov's theory of turbulence, in short. In 1922 Luis Fry Richardson (1881–1953) coined the first description of turbulence as a multiple-scale phenomenon, the so-called *Richardson cascade* [3] in which mechanical energy is injected into large eddies. Thus, turbulence's largest motions, or eddies, contain the majority of the kinetic energy. They become unstable and split into smaller eddies toward which energy is transferred, continuing until viscous dissipation stops the cascade at some small scale—the scale at which the smallest turbulent eddies dissipate into heat. Therefore, the concept of energy cascade is to be found at the core of the physics of turbulence and its multi-scale nature.

In 1941 Kolmogorov suggested a quantitative statistical description of turbulence—that is, he defined it as a self-similar cascade with universal properties [4]. After that time, the mainstream in statistical turbulence modeling has been governed by Kolmogorov's ideas, whose 1941 hypotheses have had such a powerful impact on the research of turbulence that they are known as K41 phenomenology. The power of the K41 model is its capacity to predict the universal scaling laws for velocity increments statistics. Despite its enormous success, K41 does not describe the significant features of real turbulent flows (for example, the intermittency phenomenon). Many theoretical attempts have been made to improve K41 phenomenology, including Kolmogorov's refined hypotheses of self-similarity proposed in 1962 (K62 phenomenology) and subsequently designed multifractal models [5].

The pillars of Kolmogorov's theory are to be found in three hypotheses combined with the experimental observations: 1. *Kolmogorov's hypothesis of local isotropy* states that when Reynolds numbers are sufficiently high, small-scale turbulent motions are statistically isotropic. 2. *Kolmogorov's first similarity hypothesis* states that in every turbulent flow for a sufficiently high Re, the statistics of small-scale motions has a universal form that is uniquely determined by the dissipation rate (ϵ) and the kinematic viscosity. For the two given parameters ϵ and ν, we can form the unique length $\eta = (\nu^3/\epsilon)^{1/4}$, velocity $u_\eta = (\epsilon\nu)^{1/4}$, and time $\tau_\eta = (\nu/\epsilon)^{1/2}$ scales. 3. *Kolmogorov's second similarity hypothesis*. Based on the fact that the intermediate scales l will be unaffected by the viscosity ν due to the large Reynolds number, this hypothesis states that in every turbulent flow for the appropriately high Reynolds number, the statistics of the motions of the scale l in the range $l_0 \gg l \gg \eta$ has a universal form that is uniquely determined by ϵ independently of ν. Here l_0 is the length scale, usually referred to as the *integral scale* of turbulence.

The concept of randomness. Intuitively speaking, we see randomness as the result of uncertainty in the presence of determinism. It appears as if our perception is tuned for this order in observation. It may seem counterintuitive to expect to see a "regular" picture coming from its appearance from the surrounding world of uncertainties, but it is possible. This leads us to the question: Can randomness be defined? We cannot define it formally if we apply tautological reasoning: if something is precisely defined, it is not random anymore.

The concept of a random set was first mentioned by Kolmogorov [6] in 1933. At the same time, he made the mathematical foundations of probability theory. During the 1960s, Kolmogorov put forward the theory that a binary sequence is random if its shortest description is the sequence itself [7]. While random mechanisms may exist, there is no way to establish tests that would permit one to decide whether a given data set came from a random or a nonrandom mechanism. An experiment is called *a random experiment* if its outcome cannot be predicted. Probabilistic laws govern a random process, a mathematical model of an empirical process.

We noticed the several ways to define turbulence in which an essential property is expressed verbally (moreover literally) in terms of irregularity, randomness, or chance. Usually, the degree of turbulence is considered in relation to the turbulence intensity than randomness. Stephan Pope defined randomness relative to turbulence [8]. Although his definition is reasonable and practical, it cannot be used to measure randomness because it is difficult to relate it to Kolmogorov's probability axiom.

To explain the degree of turbulence in relation to Kolmogorov complexity (KC) and its derivatives, we use Figure 8.1.

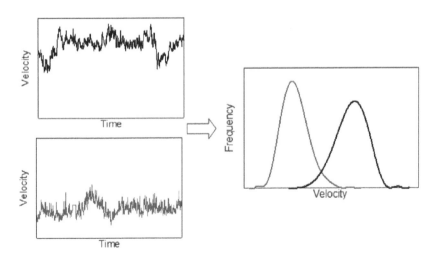

FIGURE 8.1
Toward the quantification of the randomness of turbulence in canopy flows. (Reproduced by permission from [9].)

The time series were selected among the experimental data from the turbulent flow developed in a laboratory channel with a canopy of three different densities [9]. The left-hand side of this figure depicts the two time series of the velocity measured at the two relative heights $z/k = 0.067$ (lower) and $z/k = 2.667$ (upper), respectively. The right-hand side represents the corresponding distributions of the frequencies of the velocities that occur in certain intervals (z is the vertical coordinate, while k is the canopy height). Looking at the distributions of the velocities (the right panel), we can conclude the relative height at which the turbulent motion is more random (right curve) if we use the measure of randomness based on the turbulence intensity values. As we already said in subchapter 5.5, the Kolmogorov complexity spectrum allows us to explore the range of amplitudes in a time series representing a physical system with highly enhanced stochastic components (i.e., a physical system with the highest complexity).

Figure 8.2 shows that the turbulence dynamics varies with the relative height z/k and Kolmogorov complexity spectrum's highest value. In fact, KC is higher at $z/k = 0.067$ (the left series) than KC at $z/k = 2.667$ (the right series). Consequently, it can be concluded that the turbulence regime at $z/k = 0.067$ is more random, showing the evident presence of stochastic components than at $z/k = 2.667$. In other words, these spectra can clearly point out dissimilarities between turbulent regimes for two different relative heights.

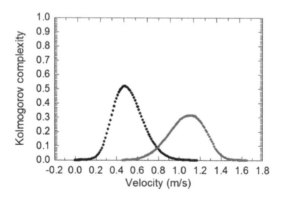

FIGURE 8.2
The KC spectra for the two time series of the velocity measured at $z/k = 0.067$ (*the left series*) and $z/k = 2.667$ (*the right series*), respectively. Both lines are fitting curves obtained from the calculated discrete values of the Kolmogorov complexity spectra. (Reproduced by permission from [9].)

8.2 The Representation of the Randomness and Complexity of Turbulent Flows with Kolmogorov Complexity

Investigations of the randomness of turbulence are accelerated scientifically and technically. A remarkable contribution to these efforts is provided by computational physics that 1. enables us to perform many computational procedures; 2. helps us to design experiments that have become more and more complex. We demonstrate an approach with KC to represent the randomness from the experiment with the two-dimensional mixing layer in turbulent flow conducted by Ichimiya and Nakamura [2]. To the reader, this subchapter may look like a technical note at first glance. However, it seemed to us that it would be a convincing illustration of the support that computational physics can provide in a problem that is no less technical than theoretical, at least in the sense that it has been seen so far. The transition to a more well-mixed state in turbulent jets is less obvious than in shear layers. This difference between shear layers and jets is interesting in the context of spatially developing flows and the evolution of the distribution of scales and turbulence spectra [10]. Ichimiya and Nakamura [2] used data in the mixing layer formed downstream of a two-dimensional nozzle. Afterward, they compressed them and represented randomness by Sato's method [11] and the approximate Kolmogorov complexity (AK) to analyze the evolution of randomness.

The description of the two-dimensional mixing layer in turbulence flow. The experimental apparatus and setup were the same as those in [12]. The mixing layer was formed at the beginning of the jet ejected into an inactive fluid from the exit ($x = 0$) of a rectangular nozzle with the following dimensions: the width (310 mm) and height h of 10 mm, as is seen in Figure 8.3. The Reynolds number based on the height h and the nozzle exit velocity U_0 were kept constant at 5000 and 7.5 ms^{-1}, respectively. Hot-wire probes with two tungsten sensing elements were used for the measurements in a range of $y \geq 0$ up to $x \leq 20$, where self-preservation is well established.

The approximated Kolmogorov complexity of the two-dimensional mixing layer in turbulence flow. In this experiment, the mean (U) and fluctuating velocities (u') of jet expansion were first measured, then the fluctuating energy's production, convection, and dissipation rates. Also, the progression of the laminar-turbulent transition in the flow-normal direction was pursued to obtain the *randomness factor* as the relationship of the energy of the continuous spectrum to the total energy in the power spectrum distribution of the velocity fluctuations [11]. After these calculations, it is possible to get a picture of the applicability of the randomness factor to the transition progress. In the part of the spectrum where the line spectral region increases in a gradual way and periodical fluctuations become dominant, the randomness factor decreases.

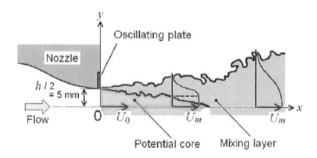

FIGURE 8.3
The diagram of the two-dimensional mixing layer and coordinate system. (Reproduced by permission from [12].)

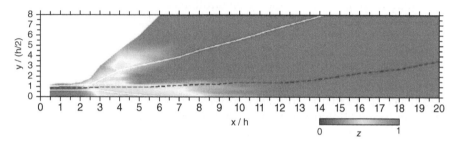

FIGURE 8.4
The isocontour map of the randomness factor z. (Reproduced by permission from [12].)

Afterward, the line spectrum changes to the continuous one—i.e., when the periodic velocity fluctuations become irregular if the randomness factor increases monotonically. In this case, a problem can occur—the randomness factor can take the same value at the two locations where the flow conditions are different [2]. Figure 8.4 shows the isocontour map of the randomness factor z in the x–y plane.

The three lines are visible on the map: two solid white lines (lower and upper) and a black dashed line. The lines are drawn in the region where the mean velocity gradient $\partial U/\partial y$ exits: 1. The positions where U/U_m is equal 0.99 near the centerline ($y/(h/2) = 0$) and 0.05 for the large y (solid white line); U_m is the velocity on the local centerline (see Figure 8.3). 2. The black dashed line is the $y/(h/2)$ position where the root mean square value of the fluctuating velocity u' becomes maximum corresponding to x. From Figure 8.4 it can be seen that after the randomness factor is once decreased in the region of $3 < y/(x/h) < 5$, it further increases downstream—that is, it does not vary monotonically.

The approximated Kolmogorov complexity of the two-dimensional mixing layer in turbulence flow. Data from hot-wire anemometers used in the velocity measurements in turbulent flow and forming the time curves are used to study the randomness of turbulence. It is inarguable that there is some elegance in theories discussing the randomness of the experimental curves of turbulent data; however, theories that directly discuss this kind of randomness are not provided by their authors. By contrast, real turbulent experimental data are discrete sequences consisting of numbers within finite time, meaning they are *par excellence* objects that should be treated with the KC of finite sequences. The randomness of an object x is defined by x-randomness in the following equation for some m: the object x is defined by m-randomness $\Leftrightarrow K(x) \geq |x| - m$.

In the analysis of data in turbulent flow, a method that employs a practical compression program is often used to estimate KC a bit length $C(x)$ of the compressed data x obtained by the chosen compression programs. Then approximated Kolmogorov complexity (AK) is the ratio of the length of the compressed data length by the program $C(x)$ and the original data length $AK(x) = C(x)/|x|$. Certainly, this is a simple approximation of complexity depending on the performance of the compression program.

Regardless of the change in the spectral distribution of turbulent flow, it is possible to determine its randomness with the analysis of complexity. Figure 8.5 shows the contour map distributions of AK in the x- and y-directions: (1) $0 < x/h < 20$ and $0 < y/(h/2)$ (Figure 8.5a) and (2) $0 < x/h < 4$ and

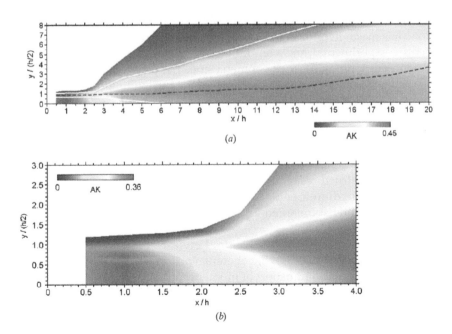

FIGURE 8.5
The isocontour maps of approximated Kolmogorov complexity (AK). (Reproduced by permission from [12].)

$0 < y/(h/2)$—i.e., the area that is just behind the nozzle exit where AK is small (Figure 8.5b).

The fact that AK takes smaller values indicates that during the compression procedure, if any regularity is detected in data, then randomness becomes small. In the region around the dashed black line, AK is large and exhibits greater complexity (Figure 8.5a). In both figures, AK is visualized, where regularity remains in the flow before and immediately after the beginning of the transition. After that region, the regularity disappears, and the flow approaches random according to transition progress. Such upstream regularity comes from the minimizing of the velocity fluctuation. Note that the randomness obtainable by data compressibility moderately decreases downstream, and AK does not necessarily change monotonically. Therefore, measuring the turbulent transition by AK alone is impossible.

8.3 The Complexity of Coherent Structures in the Turbulent Mixing Layer

Coherent structures. Turbulence in natural fluid flows and experimental conditions has much to offer in terms of the explanations of many phenomena. Recently, experimental and computational physics have remarkably contributed to describing those phenomena since the contribution of theoretical physics has become more symbolic. One of the intriguing phenomena is the appearance of a *coherent structure*, defined as the turbulent flow expressed through vortices containing orderly components and being instantaneously coherent over the spatial extent of the flow structure. Certainly, there is an expanding need to comprehend the coherent structures of a flow to consider the generation and self-sustenance of turbulence in the flow.

The coherent structures of turbulent flow (Figure 8.6) can occur at different scales and over hydraulic engineering surfaces (rivers, lakes, estuaries, open channels, including submerged aquatic vegetation, etc.), porous, and other natural surfaces.

For example, over aquatic vegetation, Kelvin-Helmholtz (KH) instability leads to the generation of large, coherent vortices within the mixing layer dominating the vertical transport of momentum through the layer. Thus, the coherent vortices downstream advected can cause the coherent waving of aquatic vegetation, referred to as *monami*. In its presence, the turbulent vertical transport of momentum is higher, with turbulent stresses penetrating an additional 30% of the plant height into the canopy [14]. The turbulent flows over the submerged aquatic vegetation surface can be considered as the analogy between aquatic flows with submerged vegetation and mixing layers due to an analogy developed for terrestrial canopies in [15].

The Representation of the Randomness and Complexity of Turbulent Flows 125

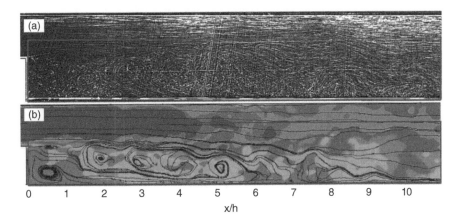

FIGURE 8.6
Coherent vortex structures in instantaneous flow: (a) visualization, (b) streamlines with velocity contours. (Reproduced by permission from [13].)

Figure 8.7 gives a schematical representation of the eddies of turbulent flow structures over and within the submerged bed roughness elements forming the roughness sublayer (RSL): 1. The RS1 zone—von Kármán vortex streets; 2. The RS2 zone is a superposition of attached eddies and KH waves (Figure 8.7a) formed around the inflection point on the mean velocity profile (Figure 8.7b). In this turbulent mixing region, the instability of KH causes coherent turbulent structures that travel downstream in the environmental fluid; 3. The RS3 zone with attached coherent eddies well above the vegetated layer. von Kármán vortex street, a repeating pattern of swirling vortices caused by a process known as vortex shedding, is shown in Figure 8.8. This image was made by Gary Koopmann and later appeared in *An Album of Fluid Motion*, assembled by Milton Van Dyke [17].

Undoubtedly, RSL with transport phenomena is an example *par excellence* of a physical complex system. For a long time, fluids mechanics has been aimed at better describing physical processes involved in flow-roughness elements interaction, including the interaction between RSL zones. The extent of the difficulty of understanding phenomena in fluid flows in an elucidative way is described by Gary Koopmann. While granting the permission for Figure 8.9, he says:

> One of my biggest thrills as a young scientist was to actually 'see' the beautiful structure in the wake of a simple cylinder that is also present in so many other natural phenomena. If you look at my *Journal of Fluid Mechanics* paper on the subject, you can also see how the motion of the cylinder motion organizes the wake flow to give rise to sustained and fluid excited vibrations. I spent many years trying to understand the physics behind this effect.

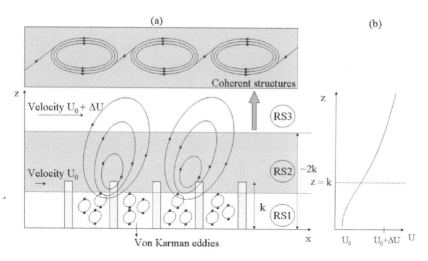

FIGURE 8.7
(a) The diagram of eddies structures over and within the canopy: 1. RS1 zone ($z/k<1$); k is the height of the cylinders used for modeling canopies where the flow field is primarily dominated by small eddies associated with the von Kármán streets; 2. RS2 zone straddles the top portion of the canopy and is dominated by a mixing layer; 3. RS3 zone ($z/k>2$) is the classical boundary-layer region dominated by eddies with length scales proportional to $(z - d)$, where d is the displacement height. (b) The mean velocity profile within the canopy $U(z) = U_0(k)\sqrt{\beta_1 e^{-\beta_3 z} + \beta_2 e^{\beta_3 z}}$ is obtained from the solution of the partial differential equation $\partial^2 u / \partial z^2 = \left[\left(2 C_d \lambda_d k_v^2\right) / \left(\sigma_s P_s\right)\right] u^2$. $U_0(k)$ is the velocity at height k; k_v is the von Kármán constant, β_1 and β_2 are parameters depending on the morphological and aerodynamic characteristics of the canopy, and $\beta_3 = \sqrt{\left(2 C_d \lambda_d k_v^2\right) / \sigma_s P_s}$; C_d—the drag coefficient; σ_s—the parameter of proportionality between the turbulent transport coefficient and velocity within the canopy; λ_d—the roughness density and P_s—the shelter factor [16]. (Reproduced by permission from [9].)

FIGURE 8.8
von Kármán vortex streets. (Figure courtesy of Gary Koopmann.)

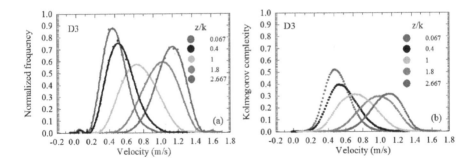

FIGURE 8.9
The distributions of velocity frequencies (a) and Kolmogorov complexity spectra (b) for the five relative flow depths (z/k). The results from an experimental study carried out in a laboratory channel with variable bed slope at the University of Naples Federico II (Naples, Italy) were used for the calculations [9]. D3 is the highest canopy density in this experiment. (Reproduced by permission from [9].)

How, then, can we reach the facts that describe the observed phenomena in fluid complex systems, particularly in turbulent flow? As mentioned above, it seems that computational physics is a good recommendation as an intermediate branch between theoretical and experimental physics (see subchapter 1.3) which bridges them with applied mathematics, computer science, and information theory as its research topic.

The profile of the complexity of turbulence in the roughness sublayer. In subchapter 8.1, we considered the Kolmogorov complexity spectrum as a measure of the randomness of the turbulent flow at two different relative heights (z/k) where the values of velocity (u) were measured. It seems that neither the conventional measure for turbulence in canopy flow nor Shannon entropy can give quantitative information about the degree of randomness. Both measures inform us either about the intensity of the turbulence in the canopy (σ_u) or about dissimilarities between amplitudes in time series obtained by measurements. Usually, in analysis of experiments with fluids, the intensity of turbulence is normalized as follows $\sigma_u = \bar{u}/u_*$, $\bar{u} = \left(\sum_{N}^{i=1} u_i^2 / N \right)^{1/2}$, where u_* is the friction velocity, while N is the number of measured velocity values (270 000 in this experiment). Figure 8.9a shows the distributions of the velocity frequencies for the selected relative flow depths for D3 canopy density (the highest density in the experiment [9]). If we plot the graph σ_u versus z/k [9], then regarding randomness, we can only say that its highest value is expected for velocities around the peak of their distributions (Figure 8.9a). Therefore, the distribution of velocity frequencies, neither quantitively nor descriptively, cannot be conclusive evidence so that we may reliably say which velocity distribution represents low or high randomness. However, the Kolmogorov complexity spectrum's highest value contains this information quantitatively expressing the randomness of turbulence in the roughness sublayer. From Figure 8.9b, randomness is higher for the eddies having lower velocities. The same figure shows that the similarities between the velocity and Kolmogorov complexity spectra differ up to the constant, which is an interesting feature. The question is whether this fact comes from physical sources or the mathematical background of the algorithm applied. Perhaps this feature is just a characteristic of these experiments, but it remains to be analyzed deeper.

8.4 Information Measures Describing the River Flow as a Complex Natural Fluid System

The turbulent flow of rivers. In subchapter 6.4, we mentioned Aksentijevic-Gibson's measure applied to streamflow time series obtained by measurements. Rivers are the part of complex natural fluid systems and, like many natural phenomena, such as earthquakes, floods, precipitation, shortwave solar radiation, and most geophysical events, are studied with time series. This approach allows us to explore their embedded dynamical structures at different time scales. At this point, we will add some details related to the nature of river flows.

When we look at a wide river in the lowland, it seems to us that it is calm and not turbulent. However, this is only the impression caused by our perception which "suggests" that the rivers' flow is not turbulent, but it is still turbulent. Recommended literature can be the paper by Björn Birnir, who theoretically proved the existence of solutions describing turbulent flow in rivers and also including an invariant measure that describes the statistical properties of that one-dimensional turbulence [18]. The Reynolds number is often used to characterize turbulent flow in rivers and streams. This number for rivers (Re_{riv}) is calculated as $Re_{riv} = \bar{D}\bar{V}/\nu$, where $\bar{D}$ is the average depth of the flow, $\bar{V}$ is the average velocity, and ν is the kinematic viscosity. Re_{riv} is typically large for streams and rivers, $Re_{riv} = 10^5-10^6$ [19].

Information measures used for streamflow time analysis. Data obtained from natural dynamical systems, such as observed streamflow time series, are inclined to noise; therefore, one should choose the corresponding information measures properly. The term "information measure" is often placed in different contexts, which causes certain confusion about how this measure should be used in natural, engineering, and computer sciences. We focused our attention on this issue in subchapters 5.1, 5.2, and Chapter 6. In this subchapter, we present the results of the three measures often used in the analysis of streamflow time series: Kolmogorov complexity, Shannon entropy, and the Lyapunov exponent.

Kolmogorov complexity (KC). This measure is described in subchapters 5.5, 6.1, and Appendix A. It conceptualizes a sequence's complexity in terms of its description length. Translated into the language of information technology, KC is the length of the shortest program necessary to reproduce the sequence. Since KC is incomputable, it is approximated with the Lempel-Ziv algorithm (LZA; see subchapter 6.1) or some of its variants. An attractive feature of KC calculated with LZA is that it is model-independent, meaning that it applies to all processes, independently of whether they are deterministic, stochastics, or stationary; therefore, it is suitable for the complexity analysis of river flow dynamics.

In the example that follows, river flow is classified according to the typology for mountains and other relief classes [20]: lowland (altitude < 200 m)—(L regime, VRB_S, UKR_D, BOS_D, DRI_K, and NER_Z), platforms and hills (200 < altitude < 500 m)—(H regime, UNA_B and DRI_B), and mountains (500 < altitude < 6000 m)—(M regime, BOS_R, MIL_S and NER_U). (The abbreviations for rivers are the same as in Figure 6.6.) Figure 8.10 shows the spatial distribution of KC for Bosnia and Herzegovina's rivers by the network, depending on their altitude.

It is seen that the spatial distribution of KC follows the relief of Bosnia and Herzegovina (Figure 6.6). In the north, there is a lowland, and the KC of the streamflow of those rivers is lower, while in the southeast and west, the values are higher (i.e., in mountainous areas where rivers have more turbulent flow).

Entropy. Entropy is one of the master problems in physics. It originates from the second law of thermodynamics and deals with a set of events. About a half-century after establishing statistical physics, Claude Shannon introduced information theory and entropy as a measure of uncertainty (see subchapter 5.3). This self-entropy is associated with a single event. Therefore, information and entropy are fundamentally related since entropy measures the average amount of self-entropies contributing to a system. The extension of entropy to information theory became a resourceful tool for modeling and associated problems in natural fluid systems because, as science, it relies on statistical methods and, therefore, on information theory. Entropy, in its different forms and heuristic modeling tools, gives practical solutions that are only sometimes optimal and ideal but are sufficient for immediate goals until a better approach is found. Note that some of the problems in the complex natural fluid system are compounded by the fact that entropy cannot easily access the structure of the data. However, despite the undoubted results

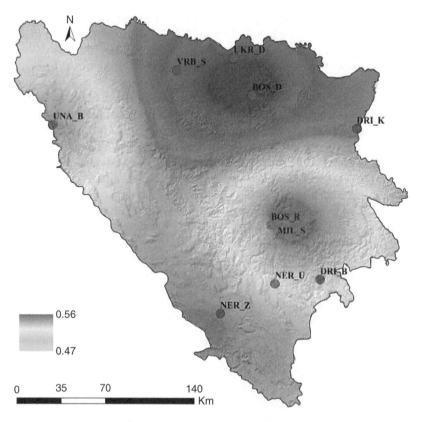

FIGURE 8.10
The dependence of the spatial distribution of KC of the monthly streamflow on altitude for seven rivers in Bosnia and Herzegovina from 1965 to 1986. (Reproduced by permission from [21].)

achieved in the application of entropy, note that combinatorial explosion at different scales makes it impossible to quantify the effect of a deep structure on finite real-time series. According to [22], the applications of entropy theory can be classified into three groups: (1) statistical or empirical, (2) physical, and (3) mixed. The first group focuses on determining probability and requires entropy maximization, including an extensive spectrum of mathematical methods. The second group derives physical relations in either time or space, while the third group is a mixture of these two.

The Lyapunov Exponent (LE). It is supposed that the average properties of a dissipative dynamical system, specifically turbulent flow, are described by a measure invariant under time evolution [23]. It is still a matter of debate whether natural fluid systems should be categorized as either deterministic or stochastic (additionally, this debate also includes discussing controversial applications of nonlinear dynamics tools). The Lyapunov exponent perhaps provides the most informative insights into a complex dynamical process. However, it belongs to informational measures that are the most difficult to determine from experimental data. This measure has one drawback—if embedding theory is used to build chaotic attractors in the reconstruction space, then the additional "spurious" Lyapunov exponents appear. Several methods are available for calculating the Lyapunov exponent, but they have to be applied carefully. For time series, it is determined as the largest Lyapunov exponent (LLE) in the spectrum of Lyapunov exponents that are organized from the largest to the smallest. LLE determines a notion of predictability for a dynamical system.

The river flow depends on many factors affecting its turbulent nature. Here we illustrate the variation of LE by using the daily streamflow time series from the twelve gauging stations on the Brazos River, whose geographical locations are shown in Figure 8.11.

The drop of the gauging stations' altitude is around 700 m. LE of the streamflow time series observed at station 1, located at the highest altitude, is 0.158, corresponding to high chaos. The streamflows observed at other gauging stations (2, 4, 5 to 12) have values of LE in the interval (0.014, 0.061) [24], which belong to the domain of weak chaos (i.e., they are very close to

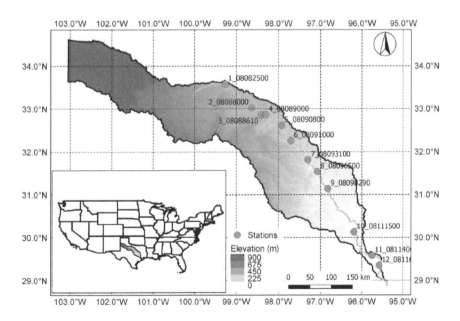

FIGURE 8.11
Geographical locations of gauging stations on the Brazos River (TX, USA). (Reproduced by permission from [24].)

zero). Surprisingly, the highest value of LE is registered for the streamflow time series observed at the gauging station 3. This jump in LE is addressed to human intervention (i.e., the presence of the Graford dam).

References

[1] J. O. Hinze, *Turbulence*, 2nd ed. New York, NY, USA: McGraw-Hill, 1975.

[2] M. Ichimiya and I. Nakamura, "Randomness representation in turbulent flows with Kolmogorov complexity (in mixing layer)," *J. Fluid Sci. Technol.*, vol. 8, no. 3, pp. 407–422, Dec. 2013, doi:10.1299/jfst.8.407.

[3] L. F. Richardson, *Weather Prediction by Numerical Process*. Cambridge, UK: Cambridge University Press, 1922.

[4] A. N. Kolmogorov, "The local structure of turbulence in incompressible viscous fluid for very large Reynolds numbers," *Dokl. Akad. Nauk*, vol. 30, pp. 301–305, 1941.

[5] M. Bourgoin, "Some aspects of Lagrangian dynamics of turbulence," in *Mixing and Dispersion in Flows Dominated by Rotation and Buoyancy* (CISM International Centre for Mechanical Sciences 580), H. J. H. Clercx and G. F. Van Heijst, Eds., New York; Berlin, Germany; Vienna, Austria: Springer-Verlag, 2018, pp. 101–127.

[6] A. N. Kolmogoroff, *Grundbegriffe der Wahrscheinlichkeitsrechnung* (Ergebnisse der Mathematik und Ihrer Grenzgebiete. 1. Folge 2). New York; Berlin, Germany; Vienna, Austria: Springer-Verlag, 1933.

[7] A. N. Kolmogorov, "On the logical foundations of information theory and probability theory," (in Russian), *Probl. Peredachi Inf.*, vol. 5, no. 3, pp. 3–7, 1969.

[8] S. B. Pope, *Turbulent Flows*. Cambridge, UK: Cambridge University Press, 2000.

[9] D. Mihailović, G. Mimić, P. Gualtieri, I. Arsenić, and C. Gualtieri, "Randomness representation of turbulence in canopy flows using Kolmogorov complexity measures," *Entropy*, vol. 19, no. 10, p. 519, Sep. 2017, doi:10.3390/e19100519.

[10] P. E. Dimotakis, "The mixing transition in turbulent flows," *J. Fluid Mech.*, vol. 409, pp. 69–98, Apr. 2000, doi:10.1017/S0022112099007946.

[11] H. Sato, "Laminar-turbulent transition in free shear flow," in *Progress in Fluid Mechanics -Turbulent Flow*. Tokyo, Japan: Maruzen, 1980, p. 75.

[12] M. Ichimiya, T. Kato, and T. Morimoto, "Effect of local periodic disturbance on mixing layer at exit of two-dimensional jet," *J. Fluid Sci. Technol.*, vol. 6, no. 6, pp. 887–901, Aug. 2011, doi:10.1299/jfst.6.887.

[13] F. Wang, A. Gao, S. Wu, S. Zhu, J. Dai, and Q. Liao, "Experimental investigation of coherent vortex structures in a backward-facing step flow," *Water*, vol. 11, no. 12, p. 2629, Dec. 2019, doi:10.3390/w11122629.

[14] M. Ghisalberti and H. M. Nepf, "Mixing layers and coherent structures in vegetated aquatic flows," *J. Geophys. Res.*, vol. 107, no. C2, pp. 3-1–3-11, Feb. 2002, doi:10.1029/2001JC000871.

[15] M. R. Raupach, J. J. Finnigan, and Y. Brunei, "Coherent eddies and turbulence in vegetation canopies: The mixing-layer analogy," *Bound. Layer Meteorol.*, vol. 78, no. 3–4, pp. 351–382, Mar. 1996, doi:10.1007/BF00120941.

[16] P. J. Sellers, Y. Mintz, Y. C. Sud, and A. Dalcher, "A simple biosphere model (SIB) for use within general circulation models," *J. Atmos. Sci.*, vol. 43, no. 6, pp. 505–531, Mar. 1986, doi:10.1175/1520-0469(1986)043<0505:ASBMFU>2.0.CO;2.

[17] M. Van Dyke, *An Album of Fluid Motion*, 10th ed. Stanford, CA, USA: Parabolic Press, 1982.

[18] B. Birnir, "Turbulent rivers," *Quart. Appl. Math.*, vol. 66, no. 3, pp. 565–594, Sep. 2008.

[19] S. L. Dingman, *Fluvial Hydrology*. New York, NY, USA: Freeman, 1984.

[20] M. Meybeck, P. Green, and C. Vörösmarty, "A new typology for mountains and other relief classes: An application to global continental water resources and population distribution," *Mt. Res. Dev.*, vol. 21, no. 1, pp. 34–45, Feb. 2001.

[21] D. Mihailović, G. Mimić, N. Drešković, and I. Arsenić, "Kolmogorov complexity based information measures applied to the analysis of different river flow regimes," *Entropy*, vol. 17, no. 5, pp. 2973–2987, May 2015, doi:10.3390/e17052973.

[22] V. P. Singh, *Entropy Theory and Its Application in Environmental and Water Engineering: Singh/Entropy Theory and Its Application in Environmental and Water Engineering*. Hoboken, NJ, USA: Wiley, 2013, p. 640.

[23] Ruelle D, "What are the measures describing turbulence?," *Prog. Theor. Phys.*, vol. 64, pp. 339–345, Feb. 1978, doi:10.1143/PTPS.64.339.

[24] D. T. Mihailović et al., "Analysis of daily streamflow complexity by Kolmogorov measures and Lyapunov exponent," *Physica A*, vol. 525, pp. 290–303, Jul. 2019, doi:10.1016/j.physa.2019.03.041.

9

The Physics of Complex Systems and Art

9.1 An Attempt to Grasp the Complexity of the Human Brain

The human brain's functioning problem is so intriguing that it is no wonder it attracted Gödel's attention [1]. He insisted on the strong separation of the mind and matter, as Wang quotes [2]:

> Many so-called philosophical problems are scientific problems, only not yet treated by scientists. One example is whether mind is separate from matter. Such problems should be discussed by philosophers before scientists are ready to discuss them, so that philosophy has as one of its functions to guide scientific research. Another function of philosophy is to study what the meaning of the world is.

Gödel approached the mind in an algorithmic manner, which was far ahead of the mainstream theories of that time. Another similar "maverick" approach occurred when, at the very end of the 1980s, Roger Penrose wrote the book *The Emperor's New Mind* [3], which was interesting and intriguing in many respects. In this book, Penrose stated that he believed that consciousness was not computational because our awareness was not a simply mechanistic derivative. We needed to alter our understanding of the physical world to take a step forward.

At the same time, the first author read in a magazine that a kind of parrot who wanted to reach a grain could surmount all the obstacles in the following experiment: Ten transparent boxes were stacked on top of each other. In the box at the top, the larger grain was placed that could pass into the box below only if the parrot managed to get it into the position for the grain to pass. All obstacles had different positions where the grain passed into the next box. Only in the tenth box could the parrot peck the grain. No matter how the arrangement of the boxes was changed, he always managed to get to the grain. For the first author, it was—and remained—a wonder how an animal brain can manifest such intelligence.

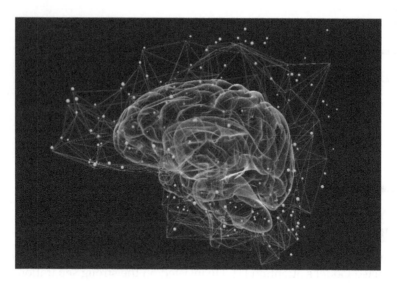

FIGURE 9.1
The human brain. (Image from the public domain.)

The prevailing opinion is that the human brain (Figure 9.1) has unique properties in the universe, but it must be the universe that we imagine and how we understand it.

According to humankind, the human brain is the most complex structure in our world. It is even more surprising that someone from the animal world, even a bird, should have such a brilliant mind. This is due to the awareness of the brain about itself because only the brain (and *perhaps* a supercomputer, according to some opinions) can show any sign of the ability to experience by *experiencing* subjectively. If we look at the recent advances in neurobiology, the mind–brain problem is still unsolved as it was a century ago. Let us suppose that science reaches one of the most outstanding achievements in its history—the complete mapping of the human brain's neurons, synapses, and neurotransmitters. This does not guarantee that we would be any closer to explaining how our thoughts and feelings are created by the brain.

Then how should we grasp this enigmatic puzzle called the brain? One possible answer is that we understand it as much as existing scientific disciplines allow us. Correspondingly, we are possibly closer to understanding it within a scientific field that currently remains undiscovered. In other words, despite the apparent progress in many sciences that deal with the brain, no existing science unifies all obtained results. Perhaps the pioneering steps of complex systems science go precisely toward that "unifying" science. Where is the place of physics in such efforts? One can often see headlines in books, magazines, or even journals, in which a scientific journalist or part of the scientific community comments how "a new theory in physics claims to solve the mystery of consciousness" or something similar. Physicists like to use a hypothesis and then make theories, but the grasp of the complexity of the human brain might require an approach we may not have mastered. We can speculate in the following way: 1 According to [4], "consciousness may be a fundamental property of nature existing outside the known laws of physics" [5], which we cannot grasp because subjective experience sometimes lies beyond scientific explanations. 2 Currently, the "tyranny" of existing physical laws does not allow us to do so (see subchapter 4.1). The prevailing opinion in the scientific world is that quantum mechanics can be considered unrelated to how the brain functions as we currently understand it. Nevertheless, Penrose's theory continues to attract the attention of the "brain"-based scientific community. Something appears to be missing in existing theories of consciousness, which may be hidden in his idea. To summarize, Penrose mentioned in an interview that he started to think about consciousness after learning about Gödel's work:

> But a powerful case can also be made that [Gödel'] results [...] established that human understanding and insight cannot be reduced to any set of computational rules. For what he appears to have shown is that no such system of rules can ever be sufficient to prove even those propositions of arithmetic whose truth is accessible, in principle, to human intuition and insight—whence human intuition and insight cannot be reduced to a set of rules.
>
> [6]

What he meant can be formulated briefly as follows: Penrose still thought that the mind was related to quantum mechanics, but it was actually to parts of the discipline that had not yet been discovered.

Let us consider the brain in the context of complex systems. It has the following properties [7]: 1 Connectivity includes interactions between individual neurons and large-scale connections between brain regions. 2 Relationships between external factors and brain (measured) outputs are nonlinear. 3 The progression of scales that range from microscopic (neuronal) to macroscopic (brain regions). 4 The brain exhibits self-organization without a centralized mechanism at both spatial and temporal scales. 5 Criticality—the brain balances between order and disorder across spatio-temporal scales. 6 Weak emergence—interactions between elements at macroscopic levels cannot be described by those at lower levels and individual parts. 7 In the brain, complex patterns occur at all scales.

The human brain has a complex structure with a multiscale organization. The brain's functioning is also complex—it is regulated by interactions between various brain units. The brain's structural organization (the brain structure) specifies its anatomical organization based on physical measures (synapses, white matter, etc.). The brain function includes interactions based on the dynamics of its elements—the activity and oscillations of individual neurons, neural ensembles, and the whole brain, along with synchronization between individual dynamics. The organization of the brain is also modular, consisting of subsystems (modules) with elements strongly linked within the same module but not so strongly connected to elements in other modules [8]. For the brain to complete cognitive tasks, modules must be synchronized while operating independently. What remains unanswered is the question of how interactions between individual brain elements and parts that result in the dynamics of the whole brain arise from its structural organization. Since numerous papers have been published on these topics, we provide a concise overview of methods from the perspective of this book: network theory, the theory of dynamical systems, and complexity measures.

Researchers have made extensive use of graph theory in neuroscience to analyze the human brain in recent years and to deal with the massive number of experimental data. Up to this point, it has been applied to study [9] (1) the structural properties of the brain through generative network models, (2) the collective behavior of neurons or ensembles (synchronization, oscillations, and avalanches), (3) how information is shared between neurons and ensembles, (4) the whole-brain regions and their interconnections, (5) some other questions, such as the understanding of cognitive control, neuromodulation, or brain dynamics in health and disease (for a more comprehensive overview see [9]). Apart from these studies, network neuroscience has also evolved and offered some new methodologies: (1) descriptive network neuroscience, (2) predictive network neuroscience, and (3) perturbative network neuroscience.

Models for brain function range from those for the activity of a single neuron to models for coupled neurons and whole-brain dynamics. The activity of a single neuron is modeled by biophysical models. The most famous model was developed by Hodgkin and Huxley [10]. This model was extended to incorporate more characteristics, resulting in hybrid Hodgkin–Huxley models, and some other models were also developed. Contrary to the causes of the single neuron activity which are primarily understood, mechanisms that underlay the collective behavior of individual neurons that leads to macroscopic activity need to be clarified. Mass neural models were developed to understand this discrepancy, consisting of several equations that model the mean activity of a group of neurons (for example, the Wilson–Cowan model [11]).

Further, individual neural mass models are coupled together to connect the dynamics of the local populations of neurons to that of distant ones—that is, to model large-scale brain activity, which led to the development of large-scale brain network models [12]. In general, dynamical models have been applied to the study of large-scale brain dynamics. However, their creators face some challenges because of difficulties in testing the predictions of these models. Chaos theory was also used; however, it was soon discovered that algorithms for detecting chaos in data do not perform well on noisy and brief empirical data [12]. In addition to the theory of dynamical systems, brain function is also studied through network and information theory. This is not discussed in this subchapter because of many papers dealing with this topic.

Finally, we discuss complexity measures in relation to the dynamics of the whole brain and some matters related to their usage. Complexity measures are most often used to differentiate between the healthy and the pathological states of the brain. Since the unique definition of the complexity of a system does not exist (and that complexity cannot be modeled), there is also no clear consensus on how the complexity of the brain is defined. Complexity is often equated to uncertainty, irregularity, or randomness. However, there is also the idea that complexity is related not to randomness but rather to the state between order and randomness [13]. Yang and Tsai [14] suggest that the complexity of the brain is its level of adaptiveness to the changing environment. From this perspective, both regular and random patterns can indicate pathology. Evidence supports this idea because both low complexity and high complexity have been observed in different neurological disorders [15]. Apart from divergent definitions, complexity measures are based on different interpretations of complexity. Even within entropy-based measures, several terms define complexity—unpredictability, regularity, or randomness [16]—while LZA approximates algorithmic complexity.

Despite their popularity in quantifying the complexity of brain dynamics, some issues are related to their usage. First, one of the most overlooked assumptions necessary for the calculation of entropy-based measures is stationarity [16]. Although the stationarity of a time series is required, they are sometimes applied to nonstationary EEG signals without any verification of this assumption. Second, different estimators and related assumptions for approximating a probability density function are used to calculate entropy-based measures; therefore, they are parameter-dependent. LZA overcomes the issues mentioned above, but it is not computable and does not apply to short sequences. These drawbacks are especially noticeable when it comes to neural time series that are recorded at high temporal resolution—LZA cannot be applied to data at smaller scales. Recently proposed measure change complexity does not require any assumptions about data and is computable and applicable to sequences of any length. Its application to the EEG data of participants with schizophrenia and healthy ones is discussed in the following chapter.

9.2 The Dualism Between Science and Art

A look at Figure 9.2 will provoke different reactions among viewers. We will speculate about only a possible few. To the majority, it probably relates to

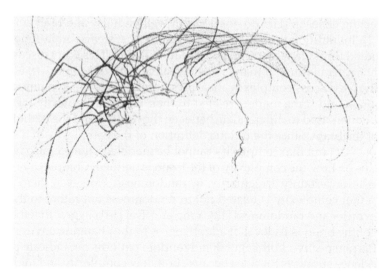

FIGURE 9.2
Salman Trtovac: Stimulated drawing. (Image courtesy of Salman Trtovac.)

something abstract and incomprehensible. Perhaps some scientists or engineers will think that this drawing belongs to the world of objects that cannot be "quantified." Some viewers (and this includes the authors of this book) will certainly accept that this quietly balanced and hence brilliant drawing is associated with the human brain, which emits the waves of its activity, or the activity of art. Why do we touch upon the relationship between science and art—that is, the relationship between physics and painting? It was—inevitable.

One would hardly expect to see the words science and art in the same sentence written in the context of their connection. But in science (and also in technology and engineering), there is a seemingly existing effort toward differentiating science from other fields that study the human brain, including art. However, we must bear in mind that both science and art come from the same source—our creativity and imagination, which points to the conclusion that it is unnecessary to make any dichotomy between them. At the most fundamental level, the endeavors of scientists and artists are intended to see the world in new ways, to grasp what they see, and then to communicate that vision. In other words, science and art manifest our human need to share knowledge and understanding of the universe.

The dilemma is whether or not art and science interact. Further, if they do, then what does this connection look like? Science and art are mostly perceived as opposites, which lack any similarities. "The other claims that art and science are only different expressions of a single underlaying voice and that similarities between the two are clues to this voice" [17]. Both attitudes seem to be based on the assumption that rationality (cognitivity) and irrationality (imagination) are regarded as exclusive opposites. What does this mean? It means that those assumptions deny the existence of the functional interdependence of rationality (cognitivity) and irrationality (imagination); they do not admit the reality of such correspondence. But the truth is that art and science are much more closely allied than we might think.

Qualities and quantities in the sciences and arts. The British novelist Charles Percy (C.P.) Snow (1905–1980) was much focused upon the dualism between science and humanism, a phenomenon which he felt "permeates and poisons the intellectual life in our times." This was highlighted by Robert Rosen in his influential book *Life Itself* [18], in which he made a step forward by saying the following:

> But the situation is, and always has been, far worse than Snow has depicted. He painted a picture of science itself as a kind of pure phase, and its relation to other aspects of our culture as a kind of phase separation; scientists and humanists separating from each other as oil separates from water, through a preference of like for like, and an antipathy of like for unlike. But the dualities that Snow depicted permeate science itself.

Therefore, in what follows we consider the dualism between science and art through the dualism between quantitative and qualitative, which is a crude approximation but sufficient for the analysis in this text.

Two quotations can serve as the most vivid illustration of the dichotomy between quality and quantity in science. One comes from the British Nobelist in physics, Ernest Rutherford (1871–1937), who remarked that "the qualitative is nothing but poor quantitative" [19]. The second is due to the American educational philosopher Robert Hutchins (1899–1977): "A social scientist is a person who counts telephone poles" [20]. When these two quotations are considered, one gets the impression that Rutherford was highly pragmatic. At the same time, Hutchins's statement is much more like a profoundly ironic comment by an English lord who had lived for a good number of years. Obviously, for Rutherford, everything placed under the umbrella of what we call quality and what we can perceive is expressible numerically—without loss or distortion.

Quality can be quantified and, therefore, it can be either measured or computed. According to Rutherford, there is no work for science until quantification is carried out. Specifically, the rough and inaccurate talk about quality has been replaced with a precise discussion about numbers. Thus, the discussion about qualities in any other terms is regarded as contemptible ("poor qualitative"). Hutchins wordlessly accepted Rutherford's crude reasoning about the connection between "science" and "quantitative," launching the phrase "social scientist," which is regarded as an oxymoron in which contradictory terms appear in conjunction. For him, the features of the qualities of social structures and the artistic worldview of the universe are precisely those that are *unquantifiable*. Contrarily, anything we can count is trivial or irrelevant (hence "telephone poles"). Rutherford's and Hutchins's attitudes are completely different. However, the issues they deal with involve the deepest levels of the relation between the observing mind and the intangible universe. Accordingly, the attitudes expressed in the quotations cannot both be right. They start from different philosophical assumptions about the nature of the observed world and the relation of the perceiver to the precepts. Essentially, that is the dualism between "hard science" and "soft science," which is personalized by Rutherford and Hutchins, respectively.

Perhaps cultural preconceptions dictate the formation of the framework for science and art. We have to bear in mind that science is analytical, while art is either intuitive or more evident—in this analysis science is logical, while art is creative. However, in synergy with other human brain activities, both contribute to improving our grasp on the universe. In subchapter 1.2, we noted that in the crisis, physics turned toward its own technologization, and therefore, for a century and more, there has been no so-called vertical discovery. Is something similar to be expected in art?

Likewise, it is an open question whether we are able to develop our own picture of the world during our life span that lasts only a few decades. One possible answer at this time would be negative. By rejecting new century values whose peak was in the epoch of Enlightenment, we are definitely in a space that does not build its own picture but can have its own metaphysics that could be found in the dialogue with the metaphysics of earlier, premodern historical epochs. This enabled the appearance of art significantly different from that of previous eras. This art is in opposition not only to the prevailing post-classical (modern + postmodern) art but also to the synthesis of classical and post-classical art. It can be defined as fundamentally new art—the last possible art [21].

9.3 Perception: Change Complexity in Psychology

In the novel *Killing Commendatore* [22], one painter gives up after realizing that he cannot paint a small waterfall landscape and the forest around it as nature made it. Following his perception, his conclusion is correct. *Perception* is our sensory experience of the world from which we attain information about the surrounding environment. Perception is the cognitive or psychological process. When an individual views a physical object based on the activity of his brain and sensory systems, then that manner of perception is *subjective perception*.

The Physics of Complex Systems and Art

In the *visual arts* (drawing, painting, sculpture, design, photography, etc.), art forms are primarily visual in nature, so perception is fundamental in creating and responding to works of art. In other words, artists should be able to recognize and understand visual phenomena and aesthetic clues. A number of key visual characteristics exist in the common perception of arts that most appeal to the human eye—symmetry, complexity, contrast, curvature, color, compactness, proportion, lines, etc.

Change complexity is an information measure initially designed to account for psychological phenomena and processes. In this subchapter, we summarize 2D change complexity and its relation to 2D visual patterns and perception that are described in [23]. R code for 2D change complexity is available in the same paper and Appendix B. Originally, psychologists were focused on discovering factors that possibly govern perception, while mathematicians joined them later. Most of those early studies put forward theories that perception gives preference to regular, simple shapes and that rules such as proximity and similarity regulate perceptual grouping. One theorist, Frederick Attneave (1919–1991), used concepts from information theory to examine perception, thereby marking the beginning of a new era in the quantitative and complexity study of perception. In several papers, he showed that some picture segments contained more information and attempted to examine the influence of matrix grain, curvedness, symmetry, number of turns, compactness, and angular variability of random-looking figures on the complexity of a figure. It was discovered that the most important factor contributing to complexity was the number of turns (changes). Subsequent studies generally reconfirmed the importance of change to the study of perception. Later, the study focused on simplicity, which established some new theories and methods in psychology, such as structural information theory and minimum description length.

The relationship between the symmetry and complexity of simple geometrical figures was examined to test the intuitive rule that less symmetrical objects are more complex, or vice versa. A square, rectangle, triangle, and scalene possess four, two, one, and zero axes of symmetry, respectively. The square is the most symmetrical, whereas the scalene has no axes of symmetry. Change complexity addressed this pattern—more symmetrical patterns were also less complex, while the loss of symmetry increased the complexity of a figure. Change complexity was further tested to examine how it performs with regard to the symmetry of random-looking patterns. Vertical symmetry, translation symmetry, rotational symmetry, and skewed symmetry were obtained by (1) rotation, (2) translation, (3) two rotations (vertical + horizontal), and (4) vertical rotation + vertical translation, and then one or more cells were shifted. In this case, change complexity also captured the intuitive order of the complexities of the figures. Vertical symmetry was the least complex, while translation symmetry had the second-lowest complexity value. The jump in the complexity of translation and rotational symmetry was significant, and, finally, the most complex was that with skewed symmetry. Perturbations also increased complexity; however, their effects were less noticeable as the form of symmetry became more complex. It should be stressed that the axis of symmetry is vertical.

Principles that govern perception may not be sufficient to explain human observations of real-world stimuli since they are often complex and dynamical. It is challenging to control stimulus and rules of perception as visual patterns become more complex, and factors such as eccentricity, symmetry, and proximity of elements fail to explain the process. Information-probabilistic concepts are used to analyze the structure of a stimulus when complexity starts to transfer from order to disorder. Change complexity was able to follow the simulated process of transition completely (Figure 9.3)—that is, complexity values increased at each step as patterns became more dissipative and less ordered.

Complexity and dynamics of the visual field. Rudolf Arnheim (1904–2007) suggested that the central position of an object and its position at the corner of a frame balance perceptual organization, while any dislocation of the object from these two points produces instability. Since these two positions are structurally different but still produce similar results, some psychologists hypothesized that the object's size could be attributed to this matter. To relate complexity to the dynamics of the visual field, Aksentijevic et al. [23] used change complexity. Simulations of the movement of a black square within a larger white square (a frame) along cardinal and diagonal axes were

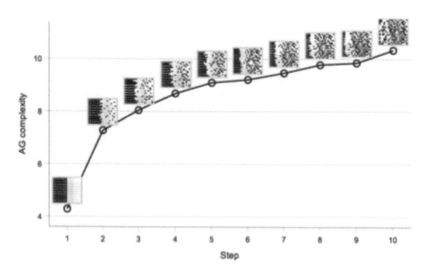

FIGURE 9.3
2D AG complexity and transition from order to disorder. (Reproduced by permission from [23].)

performed for three black squares with different sizes (6.9%, 0.2 %, and 17.4% of the total area, respectively). The complexities of all three squares were organized according to the following pattern: the lowest complexities were observed when the black square was at the corner of the white frame, and as the black square moved toward the center of the frame, complexity values increased linearly. However, at the central position of the frame, the complexity values dropped when compared with the surrounding complexity values. Also, the drops in complexity at the central position were highest for the largest square, followed by the complexities of the second-largest square. Eventually, they were the lowest for the smallest square. An explanation of this complexity pattern was based on the dynamics of the visual field characterized by rubber sheet geometry. The perceptual field can be represented by an elastic manifold. Objects distort this field, and distortion is proportional to the object's size—it is larger for larger objects and vice versa. Because of their size, large objects are sufficient for producing stability, and their movements cause the surrounding field to distort significantly.

On the other hand, the stability under tension between the field and the object's size is characteristic of small objects. Because of their size, the distortion of the surrounding field is smaller. Further, edges and corners have a natural attractive feature, and the tension in the field attracts objects toward them. Overall, the stability of the central position (related to the interaction between the object and the field) is qualitatively different from that of edges and corners (related to the field). Therefore, the lower but different complexity values characterized both locations at which stability was achieved, while increased complexity characterized the movements from these points that caused instability. Likewise, the differences in drops of complexities reflected the distortion level relative to the object's size at the central position.

Complexity and grouping of triangles. The movement of two triangles that approach each other with the bases and vertices was simulated, starting from the largest distance (both triangles were at the edge of the frame that gradually became smaller). The complexities at the starting point were similar and the lowest, with almost no difference between the two scenarios. As the distance between the triangles was smaller, the complexity values increased linearly (Figure 9.4). However, the complexity of the "base" scenario declined at the nearest locations. At the more distant points, by contrast, there was no connection between the triangles, which resulted in the lowest complexity values.

The subsequent increased complexity arose because the triangles started interacting with the field. Finally, the drop in the complexity value for the "base" scenario was due to the grouping of two triangles, which is considered more compact (relative to perception) when they face each other with their bases than when triangles face each other with their vertices. Another simulation in which triangles were smaller relative to the size of the frame mostly confirmed this result.

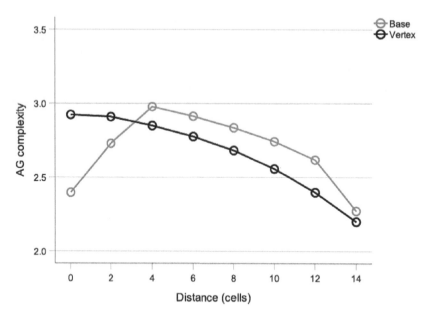

FIGURE 9.4
AG complexity versus the mutual distance between triangles. (Reproduced by permission from [23].)

9.4 Entropy, Change Complexity, and Kolmogorov Complexity in Observing Differences in Painting

In subchapter 9.2, we noted that science and art are much more closely connected than we might think. The potential closeness of that connection is demonstrated in the book *Lumen Naturae: Visions of the Abstract in Art and Mathematics* [24] by the mathematical physicist Matilde Marcolli. In this book, she explores modern art and modern science themes—the concept of space, the notation of randomness, the shape of the cosmos, etc.—by visualizing unexpected interdependencies that illuminate the universe. The concept of our book and the available space do not allow us to do something similar to Marcolli's work but with another approach. However, we demonstrate how they can be used to create a view of the painting and its history with different information measures. The artistic community is gradually relying on the automatic computational analysis for authenticating or classifying artistic paintings [25]. Indeed, the use of information measures does not mean that it is possible to make such quantification and give the value judgment of an artwork, whether it is perceived as attractive or repulsive.

In science and art, entropy is often utilized as a measure of disorder in, or lack of information about, any system. Over time, it has slowly lost a metaphorical meaning, which is perhaps described with filigree precision in the following quote:

> Even this cursory overview of uses of the concept of entropy reveals a number of ways the thermodynamic construct has served as a metaphor or an inspiration in fields far from its 'home.' While questions have been raised about the appropriateness of the transplantation, entropy seems to have become part of the intellectual landscape in some of its new environments.
>
> [26]

For example, the art theorist Rudolf Arnheim, in his famous essay *Entropy and Art: An essay on Disorder and Order*, accentuates the ideas of "entropy = disorder" and "information = order" [27]. In this essay, he tries to discover the disturbing contradiction between the aspiration for order in nature and in man in the context of entropy, which is implicit in the second law of thermodynamics.

Currently, it is possible to compute permutation entropy and statistical complexity [28] for paintings of different artistic styles. Those paintings have a different average for both metrics, allowing a hierarchical organization and clustering of styles. Figure 9.5 shows artworks' evolution through art history [29]. The artistic styles were grouped based on computation in the entropy (H)–complexity (C) plane in the following way: the computed measures were mapped onto the scale of order–disorder and simplicity–complexity, locally reflecting qualitative categories proposed by art historians.

Malevich's "Black Square." We demonstrate the use of AG complexity in the spatial complexity analysis of the famous painting "Black Square" by Kazimir Malevich (1879–1935), relying on [30]. "Black Square" consists of a centered black square occupying approximately 60% of a white background, which forms perfect symmetry. The visual field was made dynamic by moving the black square within the white frame, producing a subtle effect because of the dynamic relationship between the square and frame. Afterward, 2D AG complexity was computed, and the resulting complexities are shown in Figure 9.6. The central position, which is perfectly symmetrical, had lower complexity than the diagonal and lateral movements, while the lowest complexity was observed at the corner of the frame. This pattern of complexities agreed with the perceptual theory of the dynamics of the visual field—that is, the most stable locations are centrally and at the edge, while any shift from these positions produces an increase in complexity (or instability). The complexities at stable locations differ because of their qualitative differences (see subchapter 9.3 for an explanation). Further, AG

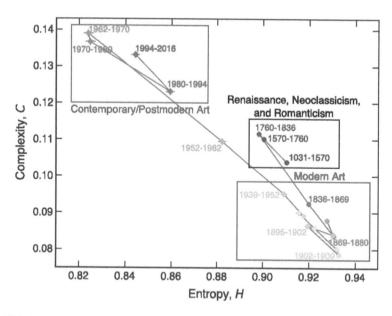

FIGURE 9.5
The evolution of artworks through the history in the complexity–entropy plane. Dots indicate the average values of H and C for a given time interval. The highlighted regions represent Renaissance, Neoclassicism, and Romanticism (*middle rectangle*), Modern Art (*lower rectangle*), and Contemporary/Postmodern Art (*upper rectangle*). (Reproduced by permission from [29].)

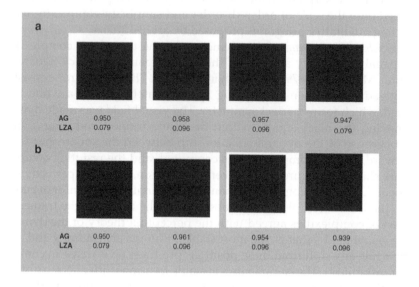

FIGURE 9.6
The influences of lateral translation (a) and diagonal translation (b) on 2D AG complexity. (Reproduced by permission from [30].)

The Physics of Complex Systems and Art

made the distinction between all locations contrary to LZA, which was only able to differentiate between two states.

Kolmogorov complexity in Orthodox iconography. In painting, the Orthodox icons of all styles are a good source for examining the existence of complex behavior and fractal patterns. However, some obstacles exist in the perception of the beauty of icons: (1) the intensity of chromatic contrasts and (2) the degree of preservation. Normalized Kolmogorov compression complexity (KC) is a good choice for overcoming these obstacles since it is insensitive to the image's size and noise in data. For example, in Figure 9.7 is seen that the Romanian icons have higher complexity than the Russian and Greek icons, which probably occurred due to the fact that the Russian icons 1. have more

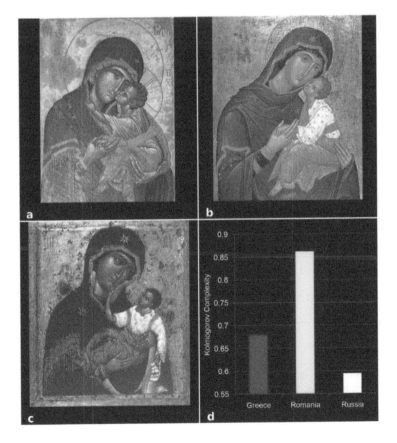

FIGURE 9.7
KC analysis of the Jesus Christ icons from the three schools of Byzantine iconographic paintings: (a) Greek, (b) Romanian, and (c) Russian school, (d) KC values. Each bar indicates a KC value of the single icon. (Reproduced by permission from [31].)

intense chromatic contrasts; and 2. are more preserved since they are made one or two centuries after the Russian and Greek icons [31].

At the end of this subchapter, we make a short elaboration on the use of Shannon entropy, AG complexity, and KC complexity in art and science because it still needs to be completely clear how values computed by these measures can be interpreted. Compared to the other two, we set an accent on KC that is only a natural but incomputable measure.

Randomness is interpreted in several ways that can be summarized as follows: (1) randomness as unpredictability, (2) randomness as typicality, and (3) randomness as complexity. Considering these interpretations, it seems that randomness is either a subjective measure or an incomputable objective measure (Kolmogorov's theory). As stated by [32], randomness is more physical than a mathematical concept—that is, it is a physical process incorporating true randomness. To calculate complexity with an algorithm, one should compute the result within a finite and practical time n. For this reason, complexity is calculated asymptotically as n approaches infinity. Note that "a mistreatment of language allows us to attribute greater randomness to a sequence with a higher complexity value" [33]. However, we maintain the term "level" in the sense of the above approach.

In our experience, authors dealing with the practical applications of information measures in art and science 1. calculate the information measure; 2. establish the changes between calculated values; and then 3. draw conclusions based on those changes. They base their discussion and explanation on the fact that the measure changes in the "rhythm" of some changes that they observed without assessing whether small changes in the measure correspond to the changes observed in the phenomenon. Indeed, this does not mean that the phenomenon is not correctly described but that the information measure is not a suitable quantifier and is only decoratively present.

References

[1] M. Dumitru, "Is human mind fully algorithmic? Remarks on Kurt Gödel's incompleteness theorems," in *Evolving Computability* (Lecture Notes in Computer Science 9136), A. Beckmann, V. Mitrana, and M. Soskova, Eds., New York; Berlin, Germany; Vienna, Austria: Springer-Verlag, 2015, pp. 23–33.

[2] H. Wang, *A Logical Journey: From Gödel to Philosophy*. Cambridge, MA, USA: MIT Press, 1996.

[3] R. Penrose, *The Emperor's New Mind: Concerning Computers, Minds, and the Laws of Physics*. London, UK: Oxford University Press, 2002.

[4] D. J. Chalmers, "Facing up to the problem of consciousness," *consc.net*. https://consc.net/papers/facing.pdf (accessed Oct. 10, 2022).

[5] S. Paulson, "Roger Penrose on why consciousness does not compute," *Nautil.us*. https://nautil.us/roger-penrose-on-why-consciousness-does-not-compute-236591/ (accessed Oct. 10, 2022).

[6] J. S. Taylor, "Gödel and human consciousness: Is the mind a closed system?," *medium.com*. https://medium.com/curious/g%C3%B6del-and-human-consciousness-46229c270f06 (accessed Oct. 12, 2022).

[7] J. Vohryzek, J. Cabral, P. Vuust, G. Deco, and M. L. Kringelbach, "Understanding brain states across spacetime informed by whole-brain modelling," *Philos. Trans. Roy. Soc. A*, vol. 380, no. 2227, Jul. 2022, Art. no. 20210247, doi:10.1098/rsta.2021.0247.

[8] D. S. Bassett and M. S. Gazzaniga, "Understanding complexity in the human brain," *Trends Cogn. Sci.*, vol. 15, no. 5, pp. 200–209, Apr. 2011, doi:10.1016/j.tics.2011.03.006.

[9] P. Srivastava, P. Fotiadis, L. Parkes, and D. S. Bassett, "The expanding horizons of network neuroscience: From description to prediction and control," *Neuroimage*, vol. 258, Sep. 2022, Art. no. 119250, doi:10.1016/j.neuroimage.2022.119250.

[10] A. L. Hodgkin and A. F. Huxley, "Propagation of electrical signals along giant nerve fibers," *Proc. Roy. Soc. B*, vol. 140, no. 899, pp. 177–183, Oct. 1952, doi:10.1098/rspb.1952.0054.

[11] H. R. Wilson and J. D. Cowan, "Excitatory and inhibitory interactions in localized populations of model neurons," *Biophys. J.*, vol. 12, no. 1, pp. 1–24, Jan. 1972, doi:10.1016/S0006-3495(72)86068-5.

[12] M. Breakspear, "Dynamic models of large-scale brain activity," *Nat. Neurosci.*, vol. 20, pp. 340–352, Mar. 2017, doi:10.1038/nn.4497.

[13] J. Ladyman, J. Lambert, and K. Wiesner, "What is a complex system?," *Eur. J. Philos. Sci.*, vol. 3, no. 1, pp. 33–67, Jan. 2013, doi:10.1007/s13194-012-0056-8.

[14] A. C. Yang and S.-J. Tsai, "Is mental illness complex? From behavior to brain," *Prog. Neuro-Psychopharmacol. Biol. Psychiatry*, vol. 45, pp. 253–257, Aug. 2013, doi:10.1016/j.pnpbp.2012.09.015.

[15] J. Ibáñez-Molina, V. Lozano, M. F. Soriano, J. I. Aznarte, C. J. Gómez-Ariza, and M. T. Bajo, "EEG multiscale complexity in schizophrenia during picture naming," *Front. Physiol.*, vol. 9, p. 1213, Sep. 2018, doi:10.3389/fphys.2018.01213.

[16] W. Xiong, L. Faes, and P. C. Ivanov, "Entropy measures, entropy estimators, and their performance in quantifying complex dynamics: Effects of artifacts, nonstationarity, and long-range correlations," *Phys. Rev. E.*, vol. 95, no. 6, Jun. 2017, Art. no. 062114, doi:10.1103/PhysRevE.95.062114.

[17] S. Richmond, "The interaction of art and science," *Leonardo*, vol. 17, no. 2, pp. 81–86, 1984, doi:10.2307/1574993.

[18] R. Rosen, *Life Itself: A Comprehensive Inquiry into the Nature, Origin, and Fabrication of Life*. New York, NY, USA: Columbia University Press, 1991.

[19] P. J. Davis, *Mathematics and Common Sense: A Case of Creative Tension*. New York, NY, USA: CRC Press, 2006.

[20] F. T. Hong, "The role of pattern recognition in creative problem solving: A case study in search of new mathematics for biology," *Prog. Biophys. Mol. Biol.*, vol. 113, no. 1, pp. 181–215, Sep. 2013, doi:10.1016/j.pbiomolbio.2013.03.017.

[21] M. Uzelac, *Filozofija poslednje umetnosti*. Novi Sad, Serbia: Veris Studio, 2010.
[22] H. Murakami, *Ubistvo Komtura: Prvi deo*, N. Tomić, Trans. Novi Sad, Serbia: Geopoetika, 2018.
[23] A. Aksentijevic, A. Mihailović, and D. T. Mihailović, "Time for change: Implementation of Aksentijevic-Gibson complexity in psychology," *Symmetry*, vol. 12, no. 6, p. 948, Jun. 2020, doi:10.3390/sym12060948.
[24] M. Marcolli, *Lumen Naturae: Visions of the Abstract in Art and Mathematics*. Cambridge, MA, USA: MIT Press, 2020.
[25] J. M. Silva, D. Pratas, R. Antunes, S. Matos, and A. J. Pinho, "Automatic analysis of artistic paintings using information-based measures," *Pattern Recognit.*, vol. 114, Jun. 2021, Art. no. 107864, doi:10.1016/j.patcog.2021.107864.
[26] C. H. Davis and D. Shaw, "The concept of entropy in the arts and humanities," *J. Libr. Inf. Sci.*, vol. 9, no. 2, pp. 135–148, 1983.
[27] R. Arnheim, *Entropy and Art: An Essay on Disorder and Order*. Berkeley, CA, USA: University of California Press, 1971.
[28] J. P. Crutchfield and K. Young, "Inferring statistical complexity," *Phys. Rev. Lett.*, vol. 63, no. 2, pp. 105–108, Jul. 1989, doi:10.1103/PhysRevLett.63.105.
[29] H. Y. D. Sigaki, M. Perc, and H. V. Ribeiro, "History of art paintings through the lens of entropy and complexity," *Proc. Natl. Acad. Sci. USA*, vol. 115, no. 37, pp. E8585–E8594, Aug. 2018, doi:10.1073/pnas.1800083115.
[30] A. Aksentijevic, D. T. Mihailović, D. Kapor, S. Crvenković, E. Nikolić-Djorić, and A. Mihailović, "Complementarity of information obtained by Kolmogorov and Aksentijevic–Gibson complexities in the analysis of binary time series," *Chaos Solitons Fractals*, vol. 130, Jan. 2020, Art. no. 109394, doi:10.1016/j.chaos.2019.109394.
[31] D. Peptenatu et al., "Kolmogorov compression complexity may differentiate different schools of Orthodox iconography," *Sci. Rep.*, vol. 12, Jun. 2022, Art. no. 10743, doi:10.1038/s41598-022-12826-w.
[32] A. Khrennikov, "Introduction to foundations of probability and randomness (for students in physics), Lectures given at the Institute of Quantum Optics and Quantum Information, Austrian Academy of Science, Lecture-1: Kolmogorov and von Mises," 2014, *arXiv*: 1410.5773.
[33] D. T. Mihailović, S. Avdić, and A. Mihailović, "Complexity measures and occurrence of the "breakpoint" in the neutron and gamma-rays time series measured with organic scintillators," *Radiat. Phys. Chem.*, vol. 184, Jul. 2021, Art. no. 109482, doi:10.1016/j.radphyschem.2021.109482.

10

The Modeling of Complex Biophysical Systems

10.1 The Role of Physics in the Modeling of the Human Body's Complex Systems

At the beginning of this subchapter, we want to emphasize that the part of the title of the chapter contains a combination of the words biophysical, complex, and systems. Authors who discuss complex systems use this combination of words and forget the role of physics in the models of any biological system that is undoubtedly a complex system starting from a cell. Biophysics is concerned with applying physical principles and methods to biological problems. Therefore, biophysical systems are complex simply in terms of the complexity of the biological system being studied (and they are always complex). At the same time, the prefix physical denotes the presence of physical phenomena in studying biological systems. We illustrate how physics is used in the models for the functioning of nanotubes in the human body. The following subchapters will focus on intercellular substance exchange and the involvement of physics and information measures to grasp autoimmune and brain diseases.

In 2004 Amin Rustom coined the term tunneling nanotube (TNT) [1] after he and his colleagues had reported *in vitro* findings of a thin structure connecting single cells over long distances. These structures allowed the transfer of membrane vesicles. The existence of the clusters of TNTs enables the formation of cellular networks with both local and long-distance interactions between cells (Figure 10.1). Empirical evidence indicates that their role is prominent in pathophysiological processes, such as the activation of natural killer cells, regulation of osteoclastogenesis, or tumor formation and growth [2]. In prokaryotes, they can be important in the process of transferring virulence from pathogenic to nonpathogenic bacteria.

The biochemical substance exchange through TNTs affects the functional stability of intercellular communication governed by gap junctions (GJs). Therefore, it is crucial to determine the threshold at which the influence

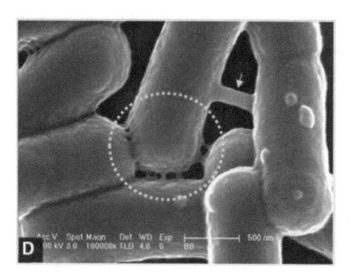

FIGURE 10.1
One illustration of intercellular TNTs between neighboring prokaryotic cells. The dashed circle indicates a cluster of smaller TNTs, while the arrow indicates a larger tube. (Reproduced by permission from [3].)

of TNTs destabilizes GJ-mediated communication. We describe this issue shortly, relying on [2, 3]. Using a simple deterministic model, we demonstrate how substance exchange through TNT affects the stability of a multicellular system. We assume that GJs create the major communication line and that the formation of TNTs can alter communication dynamics while not changing GJs. Two issues are considered in the model structure: 1. Can the transient clusters of TNTs either stabilize or destabilize intercellular communication governed by GJs? 2. The determination of the threshold at which the influence of TNTs destabilizes GJ-mediated communication [2]. In the model, GJs and TNTs allow the free passing of molecules and ions between cells. Because of the difference between anomalous and classical diffusion due to spatial inhomogeneity, it is suitable to consider the kinetics of exchange between cells through an exchange coefficient g_{ij} (in the unit of inverse time). In the simplest case, the communication between cells i and j is proportional to the concentrations between them. Therefore, the substance exchange between cells j and i can be defined as $g_{ij}(x_j(t) - x_i(t)) + \xi \delta_{ij} x_j(t)$, where $\xi > 0$ determines the strength of the influence of TNTs on the communication modeled by the uncertainty parameter δ_{ij}. Exchanged substances are involved in metabolic processes inside a cell, and they are released into the environment, which is in the model described by introducing the parameter $\alpha_i > 0$, which describes the rate by which the cell i metabolizes a substance in the time t. Since most metabolic processes follow the Michaelis–Menten dynamics, $\beta_i > 0$ is introduced as the half-time saturation coefficient for cell i. Then the intercellular communication rate can be expressed as

$$\frac{dx_i(t)}{dt} = \frac{-\alpha_i x_i(t)}{\beta_i + x_i(t)} + \sum_{j \neq i} g_{ij}\left(x_j(t) - x_i(t)\right) + \sum_{j \in N} \xi \delta_{ij} x_j(t). \qquad (10.1.1)$$

In this model, the threshold for which a multicellular system remains stable despite the influence of TNTs is determined by computing the distance to instability with the nonconvex optimization algorithm from [4]. Then humbler lower bounds are derived numerically based on pseudospectral localizations [5]. To illustrate how the concept of a model is used, several realistic scenarios were considered based on a 100-cell network, similarly as in eukaryotic tissues [6]. In all numerical tests, the lower bound of the threshold for the influence of TNTs in the infinity norm (ω_∞), the exact value in the Euclidean norm $\upsilon_{\hat{A}}^-$, and the construct pseudospectral portrait with the transient plot were computed. The following networks were used for communication determined by GJs: (1) the spatially distributed Newman-Gastner network with the weight parameter 0.001 and (2) the Erdös-Rényi modular network [7] with ten clusters connected with the overall probability of the attachment 0.03 and proportion of links within modules 90% (Figures 10.2a–c). The critical pseudospectra of the network's Jacobian matrix of GJs ($\hat{A}$) was computed (Figures 10.3a–c). At this point, the term critical pseudospectra is defined by $\varepsilon = \upsilon_{\hat{A}}^-$, which is the computed threshold for the influence of TNTs. The shadowed area indicates the extent to which the system is sensitive to changes in cell communication determined by TNTs. The network dynamics affected by the formation of TNTs were computed as the first-order approximated behavior of Equation (2), measured in the Euclidean norm, or $\|e^{t\hat{A}}\|_2$ (Figures 10.4a–c). This is only the sketch of the model's details and outputs, while more information can be found in [3].

It is interesting what can be seen from the outputs of simulations: 1. *Idealized case* when cell-to-cell communication goes only through GJs ($\xi = 0$). After passing through the short transient interval, the system reaches stability either faster (Figures 10.2c and 10.3c) or unhurried (Figure 10.4c; this situation is represented by the solid black lines). 2. *Critical case* $\xi_{crit}(\xi = \xi_{crit})$ in which the eigenvalues of the Jacobian matrix $\hat{A}$ are moved toward marginal instability (this is the worst possible scenario). Such $\hat{\Delta}$ was constructed using suitable singular vectors [6]. From Figures 10.2c, 10.3c, and 10.4c (solid gray line), the substance exchange in the system waits to move either toward stability or toward instability (oscillating mode). 3. *Case of stability*. If $\xi < \xi_{crit}$ (here $\xi = 0.8\xi_{crit}$), then the system preserves the communication integrity (the gray dashed-dotted line in Figures 10.2c, 10.3c, and 10.4c). 4. *Case of instability*. When $\xi > \xi_{crit}$ (here $\xi = 1.2\xi_{crit}$), the system is communicationally disintegrated (the gray dashed line in Figures 10.2c, 10.3c, and 10.4c).

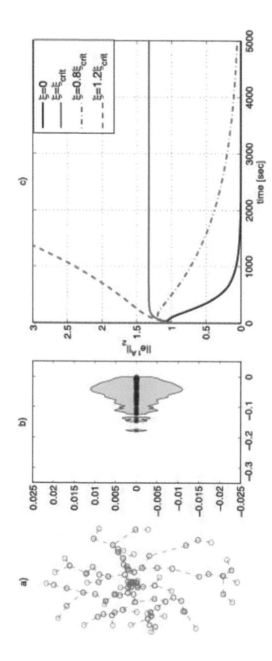

FIGURE 10.2
Computed distance to instability for 100-cell GJ network represented by the Newman–Gastner spatial network: (a) the graphical representation of the network; (b) the pseudospectral portrait of the network's Jacobian matrix A: the eigenvalues of $\hat{A}$ ($\xi = 0$) (*asterisks*); the region (*the shadowed area*) containing all possible eigenvalues for $\hat{A}$ when GJs and TNTs are included ($\xi = \xi_{crit}; \xi_{crit} = \bar{\nu}_A^-$ is a threshold value). (c) the transient growth of substance concentration within cells from the initial state for $\xi = 0$, $\xi = \xi_{crit}$, $\xi = 0.8 \xi_{crit}$ and $\xi = 1.2 \xi_{crit}$, where $\xi_{crit} = 2.38 \cdot 10^{-3}$ and $\omega_\infty = 4.47 \cdot 10^{-5}$. (Reproduced by permission from [3].)

The Modeling of Complex Biophysical Systems

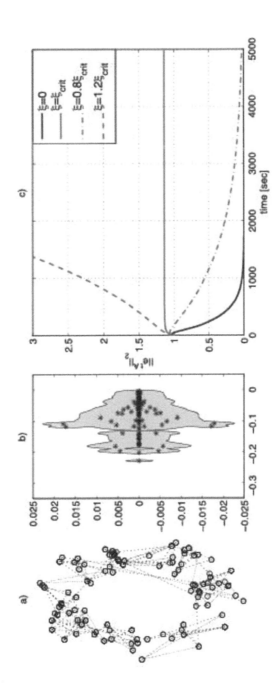

FIGURE 10.3
Computed distance to instability for 100-cell GJ network represented by the Erdös-Rényi modular network for $\xi_{crit} = 3.15 \cdot 10^{-3}$ and $\omega_\infty = 6.30 \cdot 10^{-5}$. (Reproduced by permission from [3].)

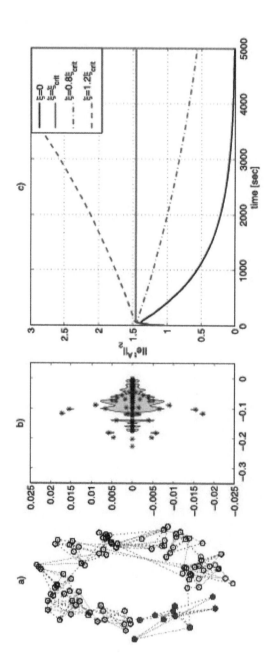

FIGURE 10.4
Computed distance to instability for 100-cell GJ network represented by the Erdős-Rényi modular network simulating a pathological state for $\xi_{crit} = 6.55 \times 10^{-4}$ and $\omega_\infty = 6.30 \times 10^{-5}$. *The filled circles*, nodes that have an altered capacity to receive signals. (Reproduced by permission from [3].)

Disrupted cell-to-cell communication can cause plentiful diseases (for example, inborn cardiac diseases). A possible influence of TNTs when communication is disrupted can be visualized by the modified Erdös-Rényi modular network with the values of parameters listed in [3]. In this network that simulates a pathological situation, the arrangement of only one module (colored circles in Figure 10.4a) exhibits cascade degradation in receiving the material under exchange; however, their capacity to send that material in the fixed network flux direction remains the same.

In the case of the disruption of only one module, the pseudospectral portrait (Figure 10.4b) depicts that the increase in the sensitivity of the whole network to changes in communication was much larger than that of a non-pathological state (Figure 10.3b). This model can also be applied successfully in FRAP (fluorescence recovering after photobleaching) to observe and quantify the rapid diffusion of calcein or other molecules between the cytoplasm of cells.

10.2 The Stability of the Synchronization of Intercellular Communication in the Tissue with the Closed Contour Arrangement of Cells

Synchronization is necessary for the proper functioning of a multicellular system disturbed by noise and internal and external perturbations. From a mathematical point of view, the stability of the synchronized state is hard to study for phase synchronization and is relatively easy to study for complete synchronization. These studies are focused primarily on complex physical systems. But for biological systems, such as tissues and organized cell assemblies, it is unclear whether the stability of the synchronized state is useful or not favorable for its functions. However, the open question "how the complexity and stability of the substance exchange process are affected by the changes in parameters that represent the influence of the environment, cell coupling and cell affinity" [6] remains unanswered.

Cube-shaped cells organized in a closed contour are shown in Figure 10.5. They are commonly found in a single layer representing a simple epithelium in glandular tissues. They are also located in the walls of tubules and the kidney and liver ducts.

We use a simple intercellular exchange model to demonstrate stability in the model of the dynamics of biochemical substance exchange in a closed contour arrangement of cells [6]. This analysis of the dynamics of different complex systems can be performed by choosing different procedures, including eigenvalue-based methods often used for asymptotic stability.

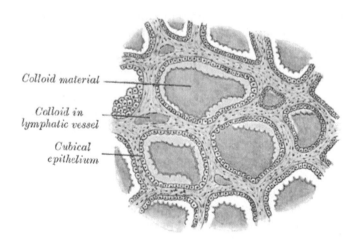

FIGURE 10.5
Cuboidal epithelial cube-shaped cells in a closed contour arrangement. (Image from the public domain.)

In the simple intercellular biochemical substance exchange model, the concentration of signaling molecules inside a cell is a main indicator of communication (Figure 10.6). Other variables and parameters of the system are as follows: 1. The cell affinity p (it formalizes an intrinsic property of a cell) by which cells perform the uptake of signaling molecules [8]. 2. The concentration c of signaling molecules in the intercellular environment. 3. The intensity of cellular response for two cells x_n and y_n in the time step n and the influence of other environmental factors r that can interfere in communication.

Hypothetically, the parameter r can represent the intracellular and intercellular environment as one variable, indicating the environment's overall disposition to communication. Further, the time evolution of the concentration in the two-cell system (x_n, y_n), $n = 0, 1, 2,...$ may be written in the form $x_{n+1} = (1 - c)\varphi$ and $y_{n+1} = (1 - c)\varphi(x_n)$, where the map $\psi : (0, 1) \to (0, 1)$ represents a cell-to-cell flow of materials. Maps $\psi(x|n)$ and $\psi(y|n)$ can be approximated by a power map where the affinity to uptake signaling molecules is indicated by the parameters p and q with the requirement that the sum of all affinities should be equal to 1, or $p + q = 1$ in the case of two cells. The system with this requirement is considered to be an isolated system in which all signals sent are received, similarly in tissues or organs. Note that the interaction between two cells in this simple model is expressed as a nonlinear coupling. The map $\varphi : (0, 1) \to (0, 1)$ models the dynamics of intracellular behavior, and it can be expressed as the logistic map $(x) = rx(1 - x)$, where r is the logistic parameter $(0 < r < 4)$ [9] pointing to the overall nature of the influence of the environment on the communication process.

The Modeling of Complex Biophysical Systems 161

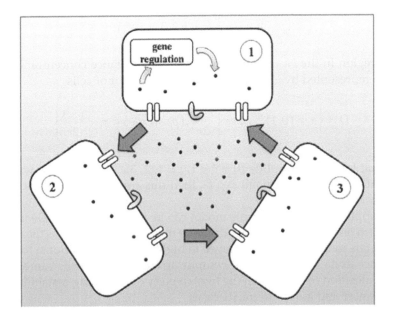

FIGURE 10.6
A schematic diagram of a model of biophysical substance exchange in a system represented with a closed contour arrangement of cells. (Reproduced by permission from [6].)

The model for N_c cells exchanging the biochemical material can be written in the form of the matrix equation

$$A = (B+F) \cdot D, \qquad (10.2.1)$$

where the matrices are defined in the following way: 1. $A = [x_{1,n+1}, x_{2,n+1}, x_{3,n+1}, \ldots, x_{Nc,n+1}]$ and $D = [x_{1,n}, x_{2,n}, x_{3,n}, \ldots, x_{Nc,n}]$ are one column matrices. 2. F is an $N_c \times N_c$ matrix h having a form similar to a bidiagonal matrix, where $f_{ij} = 0$ except for the element $f_{Nc,1} = c_{Nc} x_{1,n}^{p_{Nc}-1}$ and upper diagonal elements $f_{ij} = c_i x_{j=i+1,n}^{p_i-1}, i = 1, 2, 3, \ldots, N_c - 1$. 3. B is a diagonal matrix with diagonal elements $b_{ij} = (1 - c_i)(1 - cx_{i,n})$, $i = j$, $i = 1, 2, 3, \ldots, N_c$ and $b_{Nc,1} = 1$. Here the condition for concentrations is $\sum c_i = c$, $0 < c_i \le 1$, and x_i is the concentration of molecules in cells.

The stability of the intercellular biochemical substance exchange model. We always need to find domains for equilibrium points $\tilde{x}_i$ that permit stability for every coupling in models, such as the considered simple model. For that model, Mihailović et al. [6] find that

$$\tilde{x}_i \in \left(\frac{r-1}{2r}, \frac{r+1}{2r} \right),$$

$$(p_{i-1})^{\frac{1}{1-p_{i-1}}} < \widetilde{x}_i \, (i=1,2,3,\ldots,N_c), \tag{10.2.2}$$

or the region in the Nc- dimensional space of substance concentrations in a system represented by a closed contour arrangement of cells

$$D := \left\{ \widetilde{x}_i \in (0,1)^{N_c} : max\left\{ \frac{r-1}{2r}, (p_{i-1})^{\frac{1}{1-p_{i-1}}} \right\} < \widetilde{x}_i < \frac{r-1}{2r} \right\} \tag{10.2.3}$$

such that for every coupling ($0 < c < 1$, $i = 1, 2, 3, \ldots, N_c$, the following rule holds: if an equilibrium point $\widetilde{x}_i \in D$, then this equilibrium point is asymptotically stable.

The application of this simple model can be a source of valuable information when one wants to consider different medical issues, such as the space of parameters in the human body (even in highly chaotic conditions) for which substance exchange in a closed contour arrangement of cells can be stable and localization of changes in the human body determining instabilities that cause cancer and autoimmune diseases.

10.3 The Instability of the Synchronization of Intercellular Communication in the Tissue with a Closed Contour Arrangement of Cells: A Potential Trigger for Autoimmune Disorders

Autoimmune disorder refers to the condition in which the body attacks its tissues. Our immune system identifies foreign substances through antigens, molecules located within foreign substances and the body's tissues. In normal circumstances, only antigens from foreign substances cause the immune system to stimulate a response. Sometimes, the immune system fails to operate normally and "recognizes" the body's tissues as foreign, producing antibodies or immune cells targeting and attacking particular cells or tissues in the body. The autoimmune disorder affects tissues and organs in various ways, leading to damage and changes in their functioning (one example is the abnormal growth of an organ). We still do not know what causes autoimmune disorders. Evidence suggests that some microorganisms, drugs, and radiation in the domain of high frequencies may trigger changes confusing the immune system. This condition occurs more often in people genetically susceptible to autoimmune diseases.

In this subchapter, we point out one possibility of "confusing" the immune system in humans, which can arise due to the instability of substance exchange between cells. Using the simple model described in the previous subchapter, we demonstrate how to "find" those islands of such instability that may be a potential trigger for autoimmune diseases. This refers primarily to tissues with a closed contour arrangement of cells in the skin and glands of internal secretion.

In order to visualize the domains of stability and instability for the synchronization of intercellular communication in the tissue with a closed contour arrangement of cells, it is better to begin with studying the asymptotic stability of a two-cell system or cell-to-cell substance exchange. For that purpose, it is useful to consider the behavior of the system $v_{n+1} = F_n(v_n) \equiv L_n(v_n) + P_n(v_n)$ with the notation $L_n = [(1-c)rx_n(1-x_n)(1-c)ry_n(1-y_n)]^T$ and $P_n(v_n) = \left[cy_n^p c x_n^{1-p}\right]^T$, where $v_n = [x_n\ y_n]^T$ is a vector representing the concentration of signaling molecules inside cells [4]. Then it should be estimated whether this dynamical system can reach asymptotic stability at a possible equilibrium point $P_n(v_n)$ for two cells in the dependence of the coupling parameter c and r, including environmental factors, and p representing the intrinsic properties of the cell. The Jacobian of the map F at the point $v = \left[\tilde{x}\ \tilde{y}\right]^T$ should be computed to find the asymptotic stability regions for a two-cell system. Since the Jacobian matrix has only two eigenvalues, λ_1 and λ_2, then the discrete nonlinear dynamical system described above at the equilibrium point $v = \left[\tilde{x}\ \tilde{y}\right]^T$ is asymptotically stable for $max\{|\lambda_1|, |\lambda_2|\} < 1$ and not asymptotically stable for $max\{|\lambda_1|, |\lambda_2|\} > 1$. These conditions completely express stability and instability obtained by linearization. By contrast, if they are not satisfied, stability depends decently on the nonlinearity of the map F around the point of equilibrium v. In such circumstances, we can search the regions of stability and instability by using the eigenvalues of the Hessian matrix. Other details of a rigorous analysis of asymptotic stability are available in [2, 6]. In Figure 10.7 are drawn the regions of asymptotic stability and instability for the logistic parameter $r \in (1, 4)$ and weak coupling ($c = 0.02$). For $r = 1$ and all values of the cell affinity p (0.25, 0.50, 0.75), asymptotic stability appears in the whole domain (dark gray). Three regions can be distinguished when r increases: (1) asymptotic stability (dark gray), (2) instability (light gray), and (3) stability depending completely on the nonlinearity of the map F (white). Thus, the square section related to stability becomes smaller when r increases, which is expected in the case of the weak coupling because we move towards the region with chaotic fluctuations.

From the condition (10.2.2), it follows that the asymptotic stability of the equilibrium point is always guaranteed in the region D (10.2.3) despite the coupling parameters of individual cells. If we look at Figure 10.8 ($c = 0.6$; the strong coupling), it can be seen that the coupling parameter modifies the square region [0.375, 0625] × [0.375, 0625] but still resides in the stability region. Note that the stability of the biochemical exchange processes of a closed multicellular system can be performed by calculating the Lyapunov exponent of that system.

164   Physics of Complex Systems

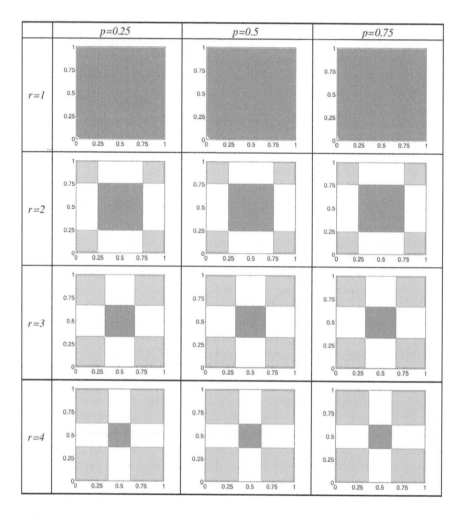

FIGURE 10.7
The regions of asymptotic stability for the equilibrium points (the conditions of asymptotic stability (1) and (2)) in the domain of the map F given in the text for the weak coupling c = 0.02. The values of the equilibrium concentration of the substance in the first cell (x) and the second one (y) are given on the horizontal and vertical axes, respectively. The asymptotic stability and instability domains are indicated by dark and light gray areas, respectively. The white area indicates that stability purely depends on the nonlinearity of the map F around the point of equilibrium v. (Reproduced by permission from [6].)

The Modeling of Complex Biophysical Systems

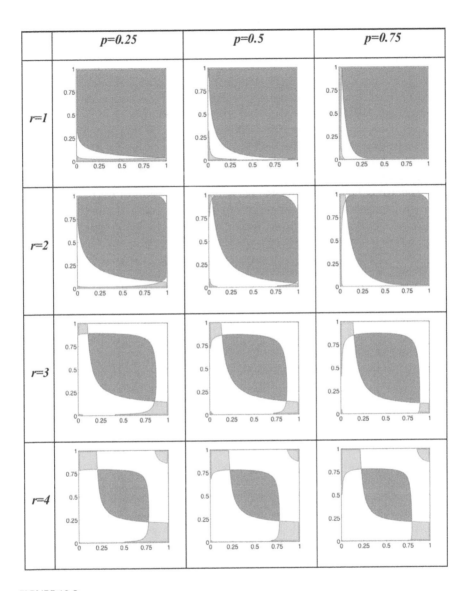

FIGURE 10.8
The same as Figure 10.7 but for the strong coupling c = 0.6. (Reproduced by permission from [6].)

10.4 The Search for Information in Brain Disorders

In the previous subchapter, we demonstrated a simple model that can be used to find the cause of autoimmune disease and perhaps even more so in its diagnosis. We already mentioned that the modeling of biophysical systems is often a one-way direction with no feedback. Perhaps the main reason for their limitations lies in the fact that no model can model the complexity of a system. In other words, we get information through the model, but it can be quite modest or insufficient.

In principle, despite a complex system, it is suitable to find such a complexity measure that is appealing to our intuition and is defined within the framework of information theory at the same time [10, 11]. A typical situation arises when looking for more detailed information because a great deal of our positive perception depends on our mind being "ready" to face it. Taking this into account, we can consider two points in searching for information in complex physical systems [2]: 1. We can determine the entropy of any system at any point in time through a measuring procedure or computation. 2. Between two successive agent's registrations in the record, there exists no information about complexity (except "that it should closely correspond to our intuition" [10])—this period is behind the window we refer to as the "black window." What information about the complexity of a system do we obtain? Complexity can be said to indicate the complexity of the transition between two states and how the corresponding entropy is computed. The only possible action is to find a suitable measure of complexity that will carry more information.

What does it look like in biology or medicine? The continuity and complexity of life processes and limited power of knowledge obtained by their observation at a certain moment and the moment of the next observation create a "black window." Therefore, it is necessary to consult experience that simultaneously takes into account several parameters: the results of measuring a process, pathophysiological knowledge of a process, outcomes from models, etc., as well as the intuition of the observer. It is interesting how medical experts met Gödel in their efforts to move frontiers in medical sciences. In Chapter 2, we elaborated on his contributions to mathematics and science, which can be summarized as follows: 1. He stated that a mathematical system possessed propositions that were impossible to prove or reject; thus, he confirmed the limited and incomplete nature of mathematical knowledge. 2. The kind of reasoning that led him to his incompleteness theorems had a great influence beyond mathematics and logic, contributing to the development of computational machines and the theory of mind, among others. Gödel's results imply some surprising things (our intuition does not encompass that) about our formal models of computability and approaches in science and medicine, particularly in immunology and psychiatry [12].

Let us note that something not covered by our intuition in science is called counterintuition, although such a term does not exist, nor is it defined in a strict philosophical sense.

Artificial intelligence systems (AI) were built, among other purposes, to help in diagnosing diseases. However, before we can trust them to handle life-and-death responsibilities, "AI will need to develop a very human trait: Admitting mistakes" [13] in the future. Up to now, AI systems have not been able to do that; in other words, this is something that computer algorithms still cannot do. It seems that this fundamental flaw is rooted in a mathematical problem. Finally, in the part that does not refer to subtle decisions that, for now, only the human mind can make, AI is necessary as a strong support for new challenges in medicine.

Here we present results in applying change complexity to brain dynamics in a case of a brain disorder—schizophrenia [14]. As mentioned earlier, complexity measures are widely used to either examine spatial brain dynamics or differentiate between the healthy and pathological states of the brain since they more appropriately reflect neural processes that exhibit complex behavior. When it comes to the complexity of brain signals, both low and high complexity can signal brain pathology. For instance, epilepsy produces low complexity. Compared with healthy participants, EEG (electroencephalogram) signals from Alzheimer's disease patients are also less complex. On the contrary, high complexity values are observed in patients with depressive disorder. This variability in the relation between complexity and pathology could be because of many variables such as age, medication, EEG recordings, data preprocessing, and various complexity measures used to address the problem. Contrary to some other mental disorders, schizophrenia is characterized by complex patterns, and no clear rules have been determined relative to the complexity and underlying mechanism of the schizophrenic brain. Since the current state of knowledge does not offer united or significant insights, the emphasis should be on analyzing information produced by the schizophrenic brain.

The Lyapunov exponent, entropy-based measures, and Kolmogorov complexity (KC) are the most important complexity measures implemented to analyze brain dynamics. However, some issues limit their application in the study of brain signals. EEGs possess low spatial resolution, and much information cannot be extracted and remain concealed. Therefore, complexity measures that are mostly used for spatial analysis cannot capture brain activity at smaller scales.

In contrast to low spatial resolution, EEG signals possess high temporal resolution, and thus far, no complexity measure has been able to successfully investigate the temporal complexity of brain dynamics successfully. Further, the foundations of chaos theory and information theory are different. Chaos theory examines complex systems governed by nonlinear deterministic rules without considering noise. By contrast, information theory is based on the assumption that a process is inherently random. However, the brain exhibits complex and nonlinear behavior that is neither truly random nor purely chaotic and arises from the interaction between its components across different spatial and temporal scales. EEG signals are considered to be nonstationary and possess a low signal-to-noise ratio. Some of these measures fail to conform to these conditions and require significant data preprocessing, which increases the risk of loss or corruption of original information. Entropy-based measures also require the selection of parameters; therefore, results could be affected by choice of parameters. As noted earlier, KC is incomputable and mostly approximated with the Lempel-Ziv algorithm (see subchapter 6.1) and partly addresses the patterning in data; however, it is impossible to estimate the complexity of short sequences.

Change complexity (AG complexity) is suitable for measuring complexity and discovering patterns in nonstationary, nonlinear, and irregular signals because it does not impose any theoretical limitations on the nature of a system. It is computable, requires no parameter setting, and allows for temporal analysis. The definition of complexity is straightforward and based on the amount of change (see subchapter 6.2). In EEG signal analysis, AG complexity has two advantages. First, it applies to short sequences, so it is possible to asses information at smaller scales. Second, it can detect rhythmicities, which allows the investigation of neural oscillations. EEG signals can be analyzed at spatial and temporal scales, and the temporal scale is further divided into four subscales — large, meso, small, and micro. All of these subscales reveal particular information about brain activity. Relying on [14], we illustrate the ability of change complexity to examine brain dynamics at both spatial and temporal scales with a particular focus on short EEG segments and neural oscillations.

Spatial analysis. Spatial analysis was conducted across 16 electrodes (O1, O2, P3, P4, Pz, T5, T6, C3, C4, Cz, T3, T4, F3, F4, F7, F8) whose placements are indicated in Figure 10.9.

The results unambiguously showed (Figure 10.10) a clear difference between the control and patient series and that the spatial complexity in the individuals with schizophrenia was lower relative to the healthy subjects across all electrodes. Spatial complexity decreased linearly from the posterior to the anterior areas, showing clear ordering among the electrodes. The same trend was observed in both series, and only complexity values differed. Also, the control and patient series were compared with 16 random sequences and revealed that brain signals are less complex than randomly generated sequences. In other words, brain dynamics is more regular than a random process.

Temporal analysis. Since evidence supports that prominent differences between healthy and schizophrenia patients occur in the brain's frontal area responsible for high-level cognitive processing, temporal analysis was mostly focused on the outer frontal electrodes (F7, F8). The mean running

The Modeling of Complex Biophysical Systems

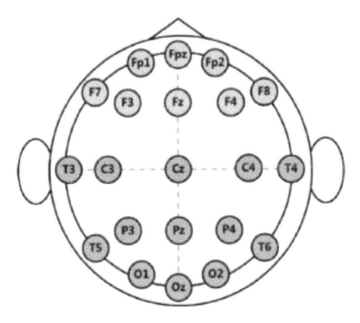

FIGURE 10.9
The 10–20 international system of EEG electrode placement: the frontal lobe (F7, Fp1, Fpz, Fp2, F8, F4, Fz, F3), the parietal lobe (C3, Cz, C4, P4, Pz, P3), the temporal lobe (T3, T4, T6, T5), and (4) the occipital lobe (O2, Oz, O1). (Reproduced by permission from [15].)

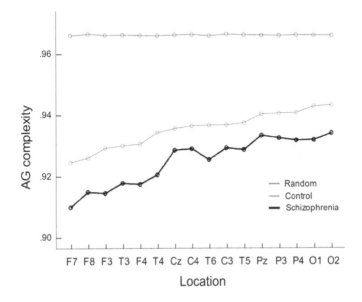

FIGURE 10.10
The mean AG complexity of EEGs across the 16 electrodes for three groups: random, control, and schizophrenia patients. (Reproduced by permission from [14].)

AG complexities recorded at F7 of the three series (Figure 10.11)—healthy, schizophrenia, random—were compared. Equal to the results of spatial analysis, the series of the healthy and schizophrenia patients were significantly less complex than the random series, distinguishing brain signals from a random process. Furthermore, the complexity of the control series

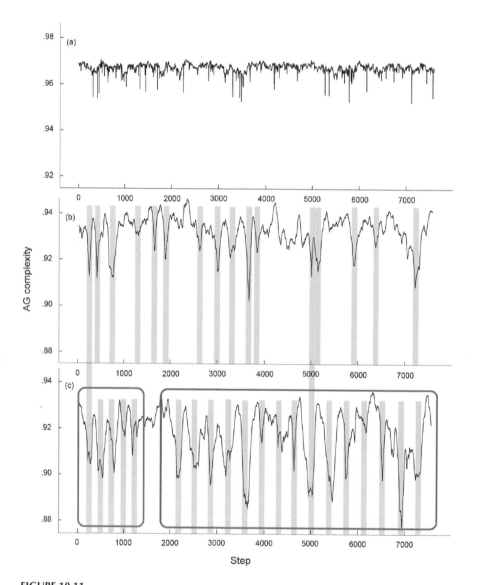

FIGURE 10.11
The running AG complexities of three groups at F7. (a): random (a), controls (b), and schizophrenia patients (c). (Reproduced by permission from [14].)

was higher than that of the schizophrenia patients, which was in agreement with the overall lower complexity in schizophrenia that resulted from spatial analysis.

Contrary to high spatial correlation, temporal correlation revealed differences between these three conditions. Periodicity was detected in the schizophrenia series, while no regularities were observed in the random and control series. The same analyses were performed at F8 and confirmed the lower complexity of the schizophrenia patients; however, only irregularly placed complexity drops without any oscillations were present in the group of individuals with schizophrenia at F8. These results—that is, reduced complexity in schizophrenia, periodicity at F7, and its absence at F8, indicated restricted activity and that pathology in schizophrenia was confined to the left frontal areas, which was also in agreement with the findings in the literature, or hypofrontality. Finally, one control and one patient with the highest complexity values were selected to search for and to compare neural oscillations. All neural oscillations were present in the healthy EEG (delta to gamma), while the same did not hold for the schizophrenia patient. This corresponded to the hypothesis of reduced higher oscillations and intensified slow oscillations (especially delta) in schizophrenia.

References

[1] A. Rustom, R. Saffrich, I. Markovic, P. Walther, and H.-H. Gerdes, "Nanotubular highways for intercellular organelle transport," *Science*, vol. 303, no. 5660, pp. 1007–1010, Feb. 2004, doi:10.1126/science.1093133.

[2] D. T. Mihailović, I. Balaž, and D. Kapor, *Time and Methods in Environmental Interfaces Modelling: Personal Insights* (Developments in Environmental Modelling 29). Amsterdam, The Netherlands; New York, NY, USA: Elsevier, 2016.

[3] D. T. Mihailović, V. R. Kostić, I. Balaž, and D. Kapor, "Computing the threshold of the influence of intercellular nanotubes on cell-to-cell communication integrity," *Chaos Solitons Fractals*, vol. 91, pp. 174–179, Oct. 2016, doi:10.1016/j.chaos.2016.06.001.

[4] V. R. Kostić, A. Miȩdlar, and J. J. Stolwijk, "On matrix nearness problems: Distance to delocalization," *SIAM J. Matrix Anal. Appl.*, vol. 36, no. 2, pp. 435–460, 2015, doi:10.1137/140963479.

[5] V. R. Kostić, L. Cvetković, and D. L. Cvetković, "Pseudospectra localizations and their applications," *Numer. Linear Algebra Appl.*, vol. 23, no. 2, pp. 356–372, Mar. 2016, doi:10.1002/nla.2028.

[6] D. T. Mihailović, V. Kostić, I. Balaž, and L. Cvetković, "Complexity and asymptotic stability in the process of biochemical substance exchange in a coupled ring of cells," *Chaos Solitons Fractals*, vol. 65, pp. 30–43, Aug. 2014, doi:10.1016/j.chaos.2014.04.008.

[7] P. Erdös and A. Rényi, "On random graphs," *Publicationes Mathematicae Debrecen*, vol. 6, pp. 290–297, 1959.

[8] D. Mihailović, I. Balaž, and I. Arsenić, "A numerical study of synchronization in the process of biochemical substance exchange in a diffusively coupled ring of cells," *Cent. Eur. J. Phys.*, vol. 11, no. 4, pp. 440–447, Apr. 2013, doi:10.2478/s11534-013-0221-5.

[9] Y.-P. Gunji and M. Kamiura, "Observational heterarchy enhancing active coupling," *Physica D*, vol. 198, no. 1–2, pp. 74–105, 2004, doi:10.1016/j.physd.2004.08.021.

[10] C. Adami and N. J. Cerf, "Physical complexity of symbolic sequences," *Physica D*, vol. 137, no. 1–2, pp. 62–69, Mar. 2000, doi:10.1016/S0167-2789(99)00179-7.

[11] D. T. Mihailović, G. Mimić, E. Nikolić-Djorić, and I. Arsenić, "Novel measures based on the Kolmogorov complexity for use in complex system behavior studies and time series analysis," *Open Phys.*, vol. 13, no. 1, pp. 1–14, 2015, doi:10.1515/phys-2015-0001.

[12] R. Tabarés-Seisdedos and J. M. Valderas, "Inverse comorbidity: the power of paradox in the advancement of science," *J. Comorb.*, vol. 3, no. 1, pp. 1–3, Mar. 2013, doi:10.15256/joc.2013.3.19.

[13] T. Haelle. "AI systems can be wrong and not admit it." *WebMD.com*. https://www.webmd.com/a-to-z-guides/news/20220405/ai-systems-can-be-wrong-and-not-admit-it (accessed Nov. 4 2022).

[14] A. Aksentijevic, A. Mihailović, and D. T. Mihailović, "A novel approach to the study of spatio-temporal brain dynamics using change-based complexity," *Appl. Math. Comput.*, vol. 410, Dec. 2021, Art. no. 126432, doi:10.1016/j.amc.2021.126432.

[15] J. H. Kim, C. M. Kim, and M.-S. Yim, "An investigation of insider threat mitigation based on EEG signal classification," *Sensors*, vol. 20, no. 21, p. 6365, Nov. 2020, doi:10.3390/s20216365.

Appendix A

A1 Lempel-Ziv Algorithm

Lempel and Ziv suggested an algorithm (LZA) (see [3] in Chapter 6) for calculating the Kolmogorov complexity (KC) of a time series $X = (x_1, x_2, x_3, \ldots, x_N)$, in which the following steps are defined: 1. Creating a sequence S of the characters 0 and 1 written as $s(i)$, $i = 1, 2, \ldots, N$, according to the rule $s(i) = 0$ if $x_i < x_t$ or 1 if $x_i > x_t$, where x_t is a threshold. Usually, the average value of the time series is chosen as the threshold. 2. Calculating the complexity counter $c(N)$, where $c(N)$ is the minimum number of distinct patterns in a given character sequence. The complexity counter $c(N)$ is a function of the length of a sequence N. The value of $e(N)$ approaches an ultimate value $c(N)$ as N approaches infinity—i.e., $c(N) = O(b(N))$ and $b(N) = \log_2 N$. 3. Calculating the normalized information measure $C_k(N)$, which is defined as $C_k(N) = c(N)/b(N) = c(N)/\log_2 N$. For a nonlinear time series, $C_k(N)$ varies between 0 and 1, but it can also be larger than 1. The flowchart for calculating KC of a streamflow series $X = (x_1, x_2, x_3, \ldots, x_N)$ using LZA is shown in Figure A1.

A2 Kolmogorov Complexity Spectrum and its Highest Value

The Kolmogorov complexity of time has the following disadvantages: 1. It cannot differentiate between a random time series and the time series having different amplitude variations. 2. Binarization of the time series can result in the loss of some information about the structure of the time series. In the complexity analysis of time series, two measures are used: (1) the Kolmogorov complexity spectrum (the KC spectrum) and (2) the highest value of the KC spectrum (see [26] in Chapter 6). The flowchart in Figure A2 shows schematically how to calculate the KC spectrum $C = (c_1, c_2, c_3, \ldots, c_N)$ for a time series $X = (x_1, x_2, x_3, \ldots, x_N)$. This spectrum allows us to investigate the range of amplitudes in a highly random time series. The KC spectrum can be computationally challenging for a large number of time series samples. The highest value K_m^C is the highest value of KC complexity spectrum ($K_m^C = \max\{c_i\}$).

173

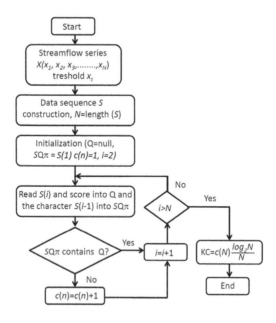

FIGURE A1
The flowchart for calculating Kolmogorov complexity using the Lempel-Ziv algorithm. (Reproduced by permission from [24] in Chapter 8.)

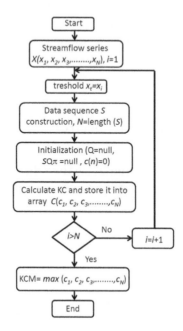

FIGURE A2
The flowchart for calculating the Kolmogorov complexity spectrum and its highest value. (Reproduced by permission from [24] in Chapter 8.)

Appendix B

The notion of Aksentijevic-Gibson (AG) complexity is new in analyzing binary time series, in which the fundamental concept underlying complexity is *change*. Change allows intuitive understanding to inform objective measures while offering a common definition of complexity across scientific disciplines. Let $A = \{0, 1\}^+$ be the set of finite nonempty binary strings over the alphabet $\{0, 1\}$. A binary string $S = s_1 s_2$ has *change* if $s_1 \neq s_2$. For $S = s_1 s_2 s_3$, we say that S has change if $s_1 s_2$ has change and $s_2 s_3$ does not and vice versa—i.e., they have different change. In general, if $= s_1 s_2 \ldots s_L$, $L \geq 2$, we say that S has change if $U = s_1 s_2 \ldots s_{L-1}$ and $V = s_2 s_3 \ldots s_L$ have different change. Here we present more mathematical descriptions of the change function, extending the material from [6] in Chapter 6.

Define an algebra $\boldsymbol{A} = (A, \cdot, [\,])$, where $A = \{0, 1\}^+$ is the concatenation of strings and $[\,] : \{0, 1\}^+ \to \{0, 1\}$ given by $[0] = [1] = 0$; $[s_1 s_2] = 0$ if and only if $s_1 = s_2$, $[s_1 s_2 \ldots s_L] = 0$ if and only if $[s_1 s_2 \ldots s_i] = [s_{L-i+1} s_{L-i+2} \ldots s_L]$, for every i, $2 \leq i \leq L$. Hence, $S = s_1 s_2 \ldots s_L$ has change—i.e., $[s_1 s_2 \ldots s_L] = 1$ if $[s_1 s_2 \ldots s_L]$ is not 0. We call $[\,]$ the *change function*.

Let $S = s_1 s_2 \ldots s_L$ and p_j be the number of substrings of S of length j which have change, $j = 2 \ldots L$. The *change profile* is the array $P = (p_2, p_3, \ldots, p_L)$. Let S be a binary string of length ≥ 2 with a change profile $P = (p_2, p_3, \ldots, p_L)$. The *complexity* C of S is

$$C = \sum_{i=2}^{L} p_i w_i, \quad w_i = \frac{1}{L - i + 1}$$

We use weights to regulate the contribution of different lengths of the substrings of S to the overall complexity of S, allowing us to model levels of structure in the complexity measure. In what follows, we concentrate on some properties of algebra $\boldsymbol{A}$.

Let t_1, t_2 be terms in the language $L^A = \{\cdot, [\,]\}$. Expressions of form $t_1 = t_2$ are called *identities* or *laws*. An identity $t_1(x_1, \ldots, x_n) = t_2(x_1, \ldots, x_n)$ holds in $\boldsymbol{A}$ or $\boldsymbol{A}$ satisfies the law $t_1(x_1, \ldots, x_n) = t_2(x_1, \ldots, x_n)$ if for every n-tuple of strings $S_1, S_2, \ldots, S_n \in \{0, 1\}^+$, $t_1(S_1, S_2, \ldots, S_n) = t_2(S_1, S_2, \ldots, S_n)$ is a true equality of $\boldsymbol{A}$. If an identity $t_1 = t_2$ holds in $\boldsymbol{A}$, we denote this fact by $\boldsymbol{A} \vDash t_1 = t_2$. The following list contains easy consequences of the definition of change function:

$$[[x]] = 0; \quad \boldsymbol{A} \vDash [[x][y]] = [[y][x]]$$

$[10^n] = 1$, $n \geq 1$ where $0^n = 00 \ldots 0$ n-times, $[01^n] = 1$, $n \geq 1$.

$$A \models [[x][y][z]] = [[z][y][x]],$$
$$A \models [[x][y[z]]] = [[[x][y]][z]],$$
$$[[x]^n] = [[y]^n]$$
$$[[x][x][y][y]] = [[x][y][y][x]].$$

However, $A \not\models [xy] = [yx]$.

Also, $A \not\models [[xy]z] = [x[yz]]$, and $A \not\models [x[y]] = [[x]y]$.

Denote by $B = (B, \cdot, [\,])$ any subalgebra of A. If $w \in B$, then $[[w]] = 0$ so that $0 \in B$ and hence $\{0\}^+ \subseteq B$. Suppose $w \notin \{0\}^+$ i.e., $w = 0^k 1\ldots$ Then $0^k 1\ldots 0^{k+1} \in B$ and $[0^k 1\ldots 0^{k+1}] = 1$ because $[0^k 1] \neq [0^{k+1}]$. Therefore, $1 \in B$ i.e., $B = A$. It follows that $(\{0\}^+, \cdot, [\,])$ is the only subalgebra of A.

2D AG complexity is a generalization of the 1D measure expressed as the mean AG complexity of all pathways (rows, columns, and diagonals) in a matrix $A = (a_{ij})_{m \times n}$. A is a binary matrix, where d is the number of diagonals of the matrix. The algorithm for 2D AG is the following ($m, n > 2$):

$$d = m + n - 1$$

$$\text{for } i = 1 \text{ to } m : r_i = AG\big((a_{i1}, a_{i2}, \ldots, a_{in})\big)$$

$$\text{for } j = 1 \text{ to } n : c_j = AG\big((a_{1j}, a_{2j}, \ldots, a_{mj})\big)$$

$$\text{for } l = 1 \text{ to } d :$$

$$\text{if} : \text{length}\big((\text{diag}_i)\big) = 1, D_i = 1$$

$$\text{else if} : \text{length}\big((\text{diag}_i)\big) = 2 :$$

$$\text{if diag}_i = (0,1) \text{ or diag}_i = (1,0), D_i = 1$$

$$\text{else } D_i = 0$$

$$\text{else } D_i = AG(\text{diag}_i)$$

$$R = \sum_{i=1}^{m} r_i, C = \sum_{i=1}^{n} c_i, D = \sum_{i=1}^{l} D_i$$

Appendix B

$$S = \frac{R}{m} + \frac{C}{n} + \frac{D}{d}$$

$$X = d - 1 + \frac{2(m-1)(n-1)}{d}$$

$$L = \frac{4mn}{3d+1}$$

$$U = (L-1)N$$

Index

Pages in *italics* refer to figures.

A

absolute
  space, 33
  time, 33
abstraction, 12, 54
Ackermann W., 21
activity
  brain, 167–168
  human, 92
  large-scale brain, 138
  restricted, 171
  single neuron, 138
  social, 4
additivity, 66
aesthetic, 143
Aksentijevic-Gibson (AG) complexity, 83–86, *144*, *146*, 147, *148–149*, 168–*169*
  running, 92
  see also Appendix B
algorithm
  Guardina-Kurchan-Lecomte-Tailleur, 74
  Lempel-Ziv (LZA), 76, 81, 168, *174*
  LZW, 81
  nonconvex optimization, 155
  nondeterministic, 30
  polynomial, 31
  RG (renormalization group) method, 102
  see also Appendix A
algorithmic
  information theory, 76, 81, 83
  randomness, 91
alphabet, 20, 86; *see also* Appendix B
Alzheimer's disease, 167
analogy, 6, 27–28, 82, 97, 124
Anderson P., 8, 99
anisotropy, 30
  of the Hamiltonian, 107
  in time, 30
antigen, 162
arrow of time, 30, 34, 52
  cosmological, 35
  psychological, 35
  quantum, 35
  radiative, 35
  thermodynamic, 34, 35
art
  physics of complex systems and, 135
artwork, 146–*148*
Aspect A., 87
asymmetry, 34–35, 52
attractor
  Hénon, 61
  Lorenz, 77
autoimmune
  diseases, 162–163
  disorders, 162
awareness, 68, 135–136
axiomatization, 74, 97

B

bacteria
  nonpathogenic, 153
  pathogenic, 153
Barish B., 87
Beckett S., 4
Bell J., 87
Bell's
  experiments, 87, 91
  inequalities, 87, 91
  theorem, 91
binarization, 76; *see also* Appendix A
bit, 67, 70, 72, 76, 123
black hole, 87–88
Black square (painting by Kazimir Malevich), 147

black swan, 3, 73
black window, 166
Boltzmann L., 67, 71–72
brain
    disorder, 167
    human, 135–138, 140, 142
    network models, 138
    see also activity
building blocks, 90

C

category theory, 11, 13–14
cell
    affinity, 159, 160, 163
    climate model grid, 60
    communication, 155, 159
    coupling, 159
    metabolic processes inside a, 154
    natural killer, 153
    prokaryotic, *154*
change complexity, 81–84, 86, 92–94, 139, 142–144, 146–*148*, *see* Aksentijevic-Gibson (AG) complexity
chaos, 12
    deterministic, 60
    in deterministic nonlinear systems, 58
    and disorder, 42
    in environmental interfaces, 58
    theory, 59, 138, 167
Chomsky N., 12
Clauser J., 87
climate
    change, 3, 42, 92
    Earth's, 41
    model grid cell, 60–61
    models, 13, 43, 50, 58, 60, 108
cognitive
    control, 138
    processing, 168
coherent structures in the turbulent mixing layer, 124–128
complex systems, 1, 9, 10–11, 14, 41–44, 49, 51–54, 56, 58, 72–76, 81–84, 92, 97, 99, 101, 103, 108–110, 127, 135–137, 153, 159, 167
    biological, 54
    modeling, 43, 51–52, 56–57
    physical, 57, 73, 75–76, 81, 100
    physics of, 14, 51, 73, 75, 82, 84, 99, 101, 135
    predictability of, 56
    universality of, 100, *see* physics
complexity, 7, 9–11; *see also* change; Kolmogorov complexity
computational
    irreducibility, 57
    physics, 12, 107, 121, 124, 127
    time, 56
    uncertainty principle, 56
consciousness, 3, 10, 68, 135, 137
contemporary/postmodern art, *148*
critical phenomena, 98–101, 103
Curie P., 52, 105

D

decoding, 7, 88
dichotomy, 54, 56, 140–141
difference equation, 39, 44, 55
differential equation, 39, 43, 50, 56–57, 59–60, 102–103, *126*
dimensional analysis, 37, 105
DNA, 72, 87
dualism, 139, 141–142
Duhem P., 11, 49–50, 52
dynamical system(s), 38, 55, 58, 60, 76, 104, 128, 131, 138–139, 163

E

earth-atmosphere
    interaction, 43
    system, 58
eddies, 118, 125–*126*, 128
EEG signals, 139, 167–168
Einstein A., 2, 5, 7, 12, 28, 36, 72, 88–90
emergence, 9, 10, 76, 98, 100
    strong, 10
    weak, 9, 10, 98, 137
emergent
    behavior, 9, 98
    phenomena, 98
    properties, 3, *10*, 98
    weakly, 98

Index

energy
 balance equation, 39, 41, *see* exchange
 free, 52, 98, 105–106
 kinetic, 110
 net, 41
 transport, 107
entangled, 87
 pairs of photons, 91
 state, 91
 system, 90
entropy, 34
 Boltzmann's, 67, 71
 Gibbs, 66
 permutation, 147
 Rényi, 66
 Shannon, 70–71, 127, 129, 150
 Tsallis, 66
environmental interface, 39, 41, 60, 111
 temperature, 39–40, 41, 83
equation
 Einstein's equation that relates mass to energy, 5
 Einstein's field equations, 30, 36, 88
 evolution, 42, 55
 of motion, 107
 Navier-Stokes, 60, 117
 nonlinear, 50, *see* energy
event, 28, 36–37, 66
 climate, 73
 ENSO (El Niño-Southern Oscillation), 84
 extremely rare, 3, 68, 73
 random, 6, 74
 rare, 73
exchange
 biochemical substance, 44, *45*, 153, 159–163
 horizontal fluxes, 58
 coefficient, 154

F

Feigenbaum M.J., 104
flow
 dissipative hydrodynamics, 59
 fluid, 82
 laminar, 117–118, 121
 renormalization, 103
 river, 92–94, 128–132
 turbulent, 117–124
Fromm E., 5, 6

G

gap junctions (GJs), 153
Gell-Mann M., 102
Gennes P.-G. De, 108
Gödel K., 2, 15, 19, 20–22, 24–28, 30–31, 36, 74, 117, 135, 166
Gödel numbering, 24–25, 27
Gödel's incompleteness theorems, 2, 3, 6, 9, 12, 14, 15, 19, 22–28, 51, 73, 98
 rotating universe model, 30
 space-time, 35–36
 time, 33, 36
 world, 27–92
Goldstein J., 9
gravitational waves, 87–88

H

Hasselmann K., 41
Hawking S., 2, 35
Heisenberg W., 56, 67, 105–108
Heisenberg's uncertainty principle, 37, 73
Helmholtz H., 124
Hénon map, 61, *see* attractor
Hilbert D., 19–21
Hofstadter D., 27

I

icons, *149*–150
imagination, 7, 140–141
immune system, 162
impression, 49, 110, 117, 128, 141
information
 abstractness of, 65–67
 amount of, 66–67, 76
 concept of, 65–67
 content, 70
 inaccessible, 65, 67
 limits of decoding, 7
 measure, 84, 128, 143, 150, 153, 166, 173

physicality of, 65–67
quantum, 67, 87, 103, 108
semantic, 7
spatial, 147
structural, 143
theory, 7, 67, 70–71, 76, 81, 83, 103, 105, 127, 129, 139, 143, 166–167
total, 66
triangle of the relationships between matter, energy and, 71–72
see also Shannon information
intelligence, 135
artificial, 167
intrinsic, 1
information, 92
property of a cell, 160
symmetries, 52
intuition, 7, 27, 30, 67, 73, 90, 97, 137, 166–167
Ising E., 99, 101, 105, 106

J

Jaki S., 6

K

Kármán T. von, *125–126*
Kelvin Lord (Thomson W.), 124
Kolmogorov A. N., 56, 76–77, 81–84, 87–92, 108–111, 117–123, 128–129, 146, 149, 167
Kolmogorov complexity, 58–59, 66, 76–77, 81–84, 87–92, 108–110, 119–123, 127–129, 146, 149, 167
approximated, 121–123
running, *59*, 82–83, 108–111
spectrum, 76–77, 87–90, 120, 127–128, 173
see also Appendix A
Kolmogorov time, 56, 58–59
Kolmogorov's probability axiom, 119
Kolmogorov's theory of turbulence, 118

L

Lempel-Ziv algorithm (LZA), 76, 81–82, 129, 139, 149, 173; *see also* Appendix A

LIGO (Laser Interference Gravitational-Waves Observatory), 7, 87–90
Lindley D., 8
Lobachevsky N., 5
logistic equation, 34, 41, *59*, 82–83, 104, 109–110
parameter, *59*, 82–83, 104, 110, 160, 163
Lorentz H., 5, 28–29, 33
Lorenz E., 57, *59*, 77
Lyapunov exponent, *40*, 44, 58, 109–110, 129–131, 163, 167
Lyapunov time, 58–59

M

Mach E., 1
Manabe S., 41
Maxwell. J.C., 72
McTaggart J. M. E., 33
Mead A., 37
metaphor, 14, 26–27, 147
metaphysics, 11, 49, 67–68, 142
mind, 135–139
human, 7, 28, 166–167
Minkowski H., 5, 19–20, *29*
model, 49–53
of biophysical substance exchange, *161*
building, 43, 52, 54, 57
choice, 52
climate, 43, 60, *60*
continuous, 39, 54
discrete, 52
Heisenberg, 105–108
Ising, 99, 101, 105–106
large-scale brain activity, 138
in mathematics, *50*
of a possible universe, 36
theory, 51
Wilson–Cowan, 118
modern art, 146, *148*

N

nanotubes, 153
natural fluid, 92, 124, 128–130
neoclassicism, 146, *148*
network

Index

and information theory, 139
Erdös-Rényi modular, 155, *157–159*
neuroscience, 138
Newman-Gastner, 155–*156*
theory, 138
neuroscience, 138
Neumann J. von, 21, 71, 74
neuron, 65, 138
neurotransmitter, 136
Newton I., 33
non-equilibrium thermodynamics, 42
non-Euclidean geometry, 5
nonlinear dynamics, 44, 131
nonlinearity, 108, 117, 163–*164*
numerical
  simulation, 13, *14*
  stability, 43

P

painting, 9, 140, 143, 146–147, 149
paradox, 11, 27, 54
Parisi G., 41
particle, 5, 37, 43, 44, 53, 56, 67, 70, 73–74, 90, 106
  Brownian, *44*
  elementary, 5, 6
  physics, 8, 73
Pekić B., 3
Penrose R., 37, 135, 137
perception, 6, 21, 26, 35, 66, 118, 128, 142–145, 149
  of freedom, 6
  subjective, 142
perceptual, 143
  field, 145
  grouping, 143
  organization, 144
  theory, 147
phase
  ordered, 98
  space, 56, 59, *61*, 76
  transitions, 52, 97–100, 102–103, 105–106
philosophy, 9, 11–12, 51, 70, 103, 135
  of science, 103
  and physics, 33, 36
  of physics, 49
physical

axioms, 11, 97
laws, 1–2, 35, 37, 51, 54–55, 72, 137
phenomena, 5, 12, 43, 51, 97, 105, 153
phenomenon, 13, 54, 73, 84, 109
processes, 2, 6, 57, 125
systems, 3, 6, 11, 41, 51, 52, 55, 57, 75, 98, 153, 159, 166
theories, 6, 12, 27, 49, 51
theory, 11, 12, 27, 49, 97–98, 100, *see* complex systems
physics
  of complex systems of, 14, 49, 51, 73, 75–76, 82, 84, 99, 101, 135
  crisis of, 4, 5
  end of, 8–9, 27
  experimental, 12–127
  generality of, 1–2
  incompleteness theorems and, 19, 27
  mathematical, 7, 51, 102
  theoretical, 7, 11–13, 102, 124
  Nobel Prize in, 41–42, 87
Planck, 5
  length, 37
  time, 37–38
  scale, 37
Planck M., 5, 26, 36
Planck's
  constant (reduced), 37
  relationship, 72
Poincaré H., 5, 20, 55
point
  critical (transition), 98–99
  "breaking", 82, *83*, 97, 108–111
  tipping, 108
power law, 53, 82, 99–100
predictability, 42, 57–59, 131, 139
  of complex systems, 56
  in hydrology, 94
  limits of, 58
  of current weather forecasting and climate models, 58
prediction, 58
  horizon, 58
  long-term weather, 58
  numerical weather, 58
  probabilistic, 91
  quantum mechanical, 38, 91, 106
Prigogine I., 42

probability
  axiom, 119
  distribution, 11, 56, 70–71, 76, 91
  model, 73
  subjective, 67
  theory, 117, 119
progress
  horizontal, 4
  in science, 1
  of physics, 9, 49
  vertical, 4, 8, 26, 73
pseudospectra, 155

Q

Quantization, 36–38
quantum
  entanglement, 89, 103, 105
  field theory, 27, 51–52, 102–103
  gravity, 37, 51
  mechanics, 8, 10, 35–39, 51, 56, 90–91, 137
  theory, 5, 11, 90, 102
qubit, 67–70

R

random
  element, 34
  event, 6, 74, 121
  experiment, 119
randomness, 7, 34, 76, 82–83, 91
  algorithmic, 91
  statistical, 91
  of turbulence in fluids, 117–120
reductionism, 3
regularity, 10, 92
  nontrivial, 10
  in river flow, 94
Renaissance, *148*
Reynolds O., 117–118, 121, 128
Riemann B., 5, 12
Rosen R., 2, 20, 141
Russell B., 22, 77
Russell's paradox, 27
Rutherford E., 141–142

S

scale
  invariance, 99–101
  time, 38, 43, 58, 97, 105, 108, 110
scaling, 97–104, 109
  invariance, 100
  laws, 100, 103, 118
  time, 41, 82
  in phase transitions and critical phenomena, 97
  propagating, 98
Schrödinger E., 2
separation of scales, 97, 101
  and capabilities of the renormalization group, 101
  in complex systems, 97
Shannon C., 70–71, 76, 83, 103, 105, 127, 129, 150
Space
  Euclidean, 29, 55
  Minkowski, 29
space–time, 15, 29–30, 35–38, 88, 101
stability
  asymptotic, 159, *163–164*
  functional, 153
  numerical, 43
Stanley H.E., 99, 105
state
  chaotic, 58
  final, 30, 57
  initial, 57–59
  microscopic, 67
  macroscopic, 67
  function, 38
  quantum, 35, 91
  space, 55
  synchronized, 159
string theory, 51
structural
  analysis, 94
  information, 7, 143
  stability, 44
  organization (the brain structure), 137–138
  properties, 138
symmetry breaking, 98

Index

synapses, 136, 137
synchronization
  complete, 159
  phase, 159
system
  biological, 1, 43, 153, 159
  biophysical, 153, 166
  coupled, *61*
  driven, 76
  dynamical, 38, 55–60, 76, 104, 128, 130–131, 138–139, 163
  ecological, 101
  entangled, 90
  isolated, 34, 160
  multicellular, 44, 154–155, 159, 163
  sensory, 5, 28, 142
Szilard L., 2, 72

T

theory of relativity, 5, 26, 28–29, 67
  general, 5, 29
  special, 5, 28
thinking
  abstract, 4, 7
  pure, 7
  visual, 30
Thorn K., 87
threshold, 76, 82, 89, 108, 153–*156*, 173; *see also* Appendix A
time
  asymmetry of, 34
  complex, 42
  of complex systems, 41–43
  continuous, 38–39, 55
  dimensionless, 41
  discrete, 38–39, 55, 104
  formation, 44
  functional, 33, 44–45
  imaginary, 38
  in philosophy, 33
  in physics, 33–36
  polynomial, 30
  quantum arrow of, 35
  step, 39, 41, 43, 55, 61, 75, 82, 160
  window, 58, 110–111
time-scale separation, 43
turbulence, 117
  in fluids, 117–123
  in natural fluid flows, 117
two-state vector formalism, 35

V

von Kármán streets, *126*
von Neumann-Bernays-Gödel set theory, 74

W

wave function, 35
wave-particle duality, 67
Weiss R., 87
white matter, 137
Wigner E., 11
Wilczek F., 37
Wilson K., 102
worldline, 29–30, 35–36

Z

Zeilinger A., 87

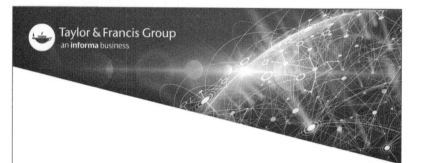